THE PATAGONIAN ICEFIELDS

Series of the Centro de Estudios Científicos

Series Editor: Claudio Teitelboim

Centro de Estudios Científicos
Valdivia, Chile

THE PATAGONIAN ICEFIELDS

A Unique Natural Laboratory
for Environmental and Climate Change Studies

Edited by

Gino Casassa

Centro de Estudios Científicos (CECS)
Valdivia, Chile

Francisco V. Sepúlveda

Centro de Estudios Científicos (CECS)
Valdivia, Chile

and

Rolf M. Sinclair

Chevy Chase, Maryland, USA

Springer Science+Business Media, LLC

Library of Congress Cataloging-in-Publication Data

The Patagonian icefields: a unique natural laboratory for environmental and climate change studies/edited by Gino Casassa, Francisco V. Sepúlveda, and Rolf M. Sinclair.
    p.   cm. — (Series of the Centro de Estudios Científicos)
    Includes bibliographical references.
    ISBN 978-0-306-46789-9   ISBN 978-1-4615-0645-4 (eBook)
    DOI 10.1007/978-1-4615-0645-4
      1. Icefields—Patagonia (Argentina and Chile) 2. Glaciers—Patagonia (Argentina and Chile) 3. Patagonia (Argentina and Chile) I. Casassa, Gino, 1958– II. Sepúlveda, Francisco V., 1948– III. Sinclair, Rolf M., 1929– IV. Series.

GB2458.5.P37 P37 2002
551.31′2′09827—dc21

                                           2002067813

Proceedings of a meeting on the Patagonian Icefields, held March 24–28, 2000, on the Chilean Naval Vessel *Aquiles*, while travelling from Punta Arenas along the west coast of Chile to Puerto Montt, and March 29, 2000, at CECS in Valdivia.

ISBN 978-0-306-46789-9

© 2002 Springer Science+Business Media New York
Originally published by Kluwer Academic / Plenum Publishers, New York in 2002

10 9 8 7 6 5 4 3 2 1

A C.I.P. record for this book is available from the Library of Congress

PREFACE

The majesty of the icefields is beyond description. He who has been fortunate to be there once, remains bound forever. To a theoretical physicist working on black holes the icefields produce a familiar vertigo, the instinctive certainty of being confronted with something so simple and beautifully extreme that it must be of importance.

The meeting whose proceedings are contained in this volume was conducted onboard of a vessel that went to the icefields, and the participants could literally set foot on them. It was expected that, for those who had not been there before, this would constitute a ritual of initiation. And so it did. For this reason we like to refer to the meeting as an expedition because, although it did not have the hardship, it had the spirit.

After this foundational expedition there have been two others, this time both with spirit and hardship, one from Chile and one from Argentina. At the moment of this writing, a fourth, full-fledged airborne expedition to the icefields is about to depart from Valdivia. Many of the people of many nations who were on board of the *Aquiles* will take part in it. We look forward to its results, and to an ongoing exciting scientific adventure.

Claudio Teitelboim
Director, Centro de Estudios Científicos

Valdivia, September 2001

ACKNOWLEDGMENTS

This volume is the result of the effort and generosity of many people and institutions. Among the institutions that made it all possible we are especially indebted to: the Millennium Science Initiative, the Tinker Foundation, the Packard Foundation, the Chilean Ministries of Defense and Foreign Affairs, the Estado Mayor de la Defensa Nacional de Chile, the Chilean Navy and Air Force and the Jefatura de Gabinete de Ministros of Argentina. Institutions are made out of people and very special people made the difference. Our warm gratitude goes to Soledad Alvear, Norma Cadoppi, Hernán Couyoumdjian, Mario Fernández, Angel Flisflisch, Eduardo Frei Ruiz-Tagle, Phillip Griffiths, Thomas Hexner, Ricardo Lagos, Heraldo Muñoz, Martha Muse, Edmundo Pérez Yoma, Fernando Rojas Vender, David Sabatini, Rodolfo Terragno, Juan Gabriel Valdés and Miguel Angel Vergara. In the midst of their high responsibilities they found time and enthusiasm to resonate to this adventure. We are especially thankful to Javier del Río of the Chilean Air Force and to Felipe Carvajal of the Chilean Navy, for their active support during the expedition/seminar. Commander Romero, his crew and the helicopter pilots of the A. P. Aquiles of the Chilean Navy provided the unusual logistics and wonderful atmosphere for the development of the meeting. We also thank Rolf Kilian for valuable assistance at his base camp in the fjords of Patagonia. We appreciate the participation of our colleagues from many parts of the world who came at short notice.

Special gratitude is expressed to Karen Everett for her help in organizing the meeting and her diligent and patient work in editing and compiling the present volume. The valuable help of Allan Rasmussen is acknowledged, who made enriching editorial comments on the papers.

INTRODUCTION

The Patagonian icefields:
A unique natural laboratory

1. INTRODUCTION

The icefields of southernmost Chile and Argentina are the largest bodies of ice in the Southern Hemisphere outside of Antarctica. In comparison to polar glaciers of the Arctic and the Antarctic, which have been intensively studied by multinational groups for many decades, the current scientific understanding of glaciers in Patagonia is still very limited.

The recent agreement on the Chilean-Argentine boundary in the Southern Patagonia Icefield, and the decision by both Governments to foster joint studies in this area, led to a meeting, held March 24-29, 2000, under the auspices of the Centro de Estudios Científicos (CECS), to assess the scientific potential of the icefields. The group of scientists assembled at the meeting is here referred to as "Icefields Scientific Task Force". This volume is a collection of papers presented at that meeting, which describes a number of aspects of the icefields, and a future major research program designed to unveil unknown fundamental scientific questions, expecting that it will contribute to a wider understanding on environmental processes in South America and globally, in addition to study subjects of great practical and timely importance to both Governments.

The icefields are known as Northern Patagonia Icefield and Southern Patagonia Icefield, extending north-south along the Andes for approximately 500 km, from 46°30' S to 51°30' S, centered on 73°30' W. The two icefields are separated by a wide, non-glaciated gap within the Andes mountains where the two major river systems of Baker and Pascua flow to the Pacific Ocean. The mean width of the icefields is 35 km, with a joint area of 17,200 km^2 and a mean altitude of the upper plateau of 1600 m, with peaks reaching 2500 m, and a few exceeding 3000 m. There are 74 major outlet glaciers and several hundred minor glaciers that drain the inland ice to the Pacific and to the large eastern piedmont lakes. The icefields, with the status of National Parks in Chile and Argentina, represent a major source of fresh water for both countries.

Most of the meeting (March 24-28) was held onboard the Chilean Naval vessel *Aquiles*, while travelling from Punta Arenas along the coast of Chile to Puerto Montt. A final meeting day was held March 29 at CECS in Valdivia. The voyage on the *Aquiles* gave us a rare chance to reach the region flanking the icefields. With the cooperation of Chilean Navy helicopters we were able to spend one day on the Pío XI Glacier.

The attendees at the meeting came principally from Chile and Argentina, but also included scientists from a number of other countries. They brought to the meeting a broad range of expertise in research in the Arctic, the Antarctic, and in studies of mountain glaciers in non-polar latitudes.

It was clear at the meeting that the icefields are largely a "blank spot" on the scientific maps of the world. Although a few systematic studies have been undertaken on the lower reaches of the glaciers, much basic research remains to be done, particularly within the upper icefield areas.

The papers of this volume are grouped under four headings: Biology, Climate and Paleoclimate, Glaciology, and Ice Coring. They were presented at the meeting to all of the attendees. Then on the last day of the meeting, four working groups discussed in detail different aspects of a proposed research program and prepared recommendations in each area.

It is evident from several of the papers that research activities are already under way in the icefields. Also, studies already carried out elsewhere, described in the rest of the papers, would have immediate application to the proposed program. Based on this work and the experience of the participants, the present meeting investigated a full range of experiments possible to perform, in order to understand the nature of the icefields, their history, and the biological and climatic records stored within them.

What follows is a summary of the recommendations of the four working groups, together with various points raised in the papers of this volume.

2. WORKING GROUPS

2.1. Biology

The icefields constitute a unique and unusual archive of climatic and biological information. They are also unusual in being in close proximity to sub-Antarctic rain forests and complex temperate ecosystems. This would make it possible to find a record of biodiversity over the last thousand years, and its correlation with climate change. This includes records of microbial diversity (DNA analyses) and higher plants (pollen analyses). These records can be compared with those from areas surrounding the icefields.

Studies would be continued on the algae and insect communities that live on the icefields, which contain an unusual group of extremophiles.

2.2. Climate and Paleoclimate

There has not been to date a systematic collection of weather records within the interior of the icefields. Because of the importance of obtaining such records in order to understand and predict the local weather patterns, this group recommended the installation of a network of automatic weather stations on the ice on several transects across the region.

The climate data that can be obtained from ice cores can be compared with that collected from other proxy records of the surrounding environment, and can be used to detect global changes and hemispheric cycles such as El Niño/La Niña.

Records of non-periodic events, such as radioactive fallout, volcanic activity, and meteorite impacts can also be preserved in the ice.

2.3. Glaciology

The glaciers of the icefields are thought to be very sensitive to climate change. They are presently in fast retreat, with a few exceptions of stable and advancing glaciers. Many basic glaciological questions remain unanswered, such as the flow dynamics, the causes driving glacier changes, the mass balance of the icefields and their contribution to global sea level.

This group recommends an intensive study of the glaciers, measuring climatic and glacial parameters within the periphery and the interior of the icefields. Such data are essential input for modeling the present ice flow, the past history, and the future evolution of the icefields.

2.4. Ice Coring

To obtain the records existing in the icefields, it will be necessary to secure ice cores from depths of 200 m or more. This group sketched a plan that would achieve within five years drilling and ice-core retrieval from such depths. At first shallow cores would be obtained at a number of sites, to evaluate the paleoenvironmental potential. Together with collection of geophysical and glaciological data, this will allow the selection of an optimal deep drilling site.

The weather conditions on the ice can be extreme (see Godoi *et al.*, Chapter 14), and are often worse than at other locations where such drilling has been performed. Drilling routinely and successfully on the icefields will require a major scientific and engineering program to adapt drilling techniques to the local conditions.

3. IN CONCLUSION

These icefields are unique on this planet, especially so within the Southern Hemisphere. They are flanked by a variety of terrestrial and marine ecosystems, which influence the biological and climate record in the ice. The icefields deserve a careful study since they are at present largely a "blank spot on the map". A major research program in the icefields of southern Chile and Argentina would be of significant value, both to answer fundamental scientific questions and to study subjects of great practical and timely importance, fostering in the process binational and international collaboration.

The ice on the whole is likely to be relatively young, probably less than one thousand years, and the high annual precipitation leads us to expect to find within the ice short high-resolution climate and biological records. All the groups agreed that the icefields should be studied in the broader context of their proximity to sub-Antarctic rainforests and to a rich temperate ecosystem. Much of the biological and climate information that can be recovered from the ice will be related to these surroundings, and will give valuable insight about the climate and the biological history of the last centuries of southernmost South America. More broadly, the icefield studies would form an important link in the "Pole-Equator-Pole" environmental studies now under way in the Americas, stretching from the Arctic to the Antarctic, filling in an information gap between tropical latitudes and the Antarctic which would be especially important for global change research.

Due to their temperate nature, the icefields are sensitive to climate changes, both local and worldwide. We must understand the present nature of the icefields and their history to be able to make predictions of their future evolution. This is particularly important since they represent a large potential for water resources and hydroelectric power for both Chile and Argentina. This will require a number of transect studies to understand the icefields in their entirety. Some of this can be done remotely by aerial and satellite imaging, but much can only be done by "ground truthing" on the ice itself.

The studies planned are inherently medium and long-term and will require multi-year planning. A facility for handling ice cores exists presently in Argentina (Mendoza), but there is none currently in Chile. There are good reasons to consider creating such facilities in Chile, and improving the already existing one in Argentina. Both the efficiency in work, personnel safety, and supporting infrastructure must be planned carefully. This may involve jointly the Military Forces of both Chile and Argentina in some stages of the fieldwork, since they have the necessary capabilities and experience already in place.

The experience and expertise among the scientists of Chile and Argentina provide the basis for the proposed studies. The glaciological and biological studies currently underway in the two countries already form a beginning for the proposed more extensive work. International collaboration will, however, be essential to harvest the secrets the ice contains for the benefit of a better understanding of our planet.

CONTENTS

GLACIOLOGY

ICE CORING

LIST OF PARTICIPANTS

Alberto Aristarain   Laboratorio de Estratigrafía Glaciar y Geoquímica del Agua y de la Nieve (LEGAN), Instituto Antártico Argentino, Mendoza, Argentina, Tel +54 261 428 8808, Fax +54 261 428 7370, email aristar@lab.cricyt.edu.ar

Carlos Bustos   Ministerio de Relaciones Exteriores, Santiago, Chile

Jorge Carrasco   Dirección Meteorológica de Chile, Santiago, Chile, Fax +56 2 6019590, email jcarrasco@meteochile.cl

Gino Casassa   Centro de Estudios Científicos (CECS), Avenida Arturo Prat 514, Casilla 1469, Valdivia, Chile, Tel +56 63 234500, Fax +56 63 234517, email gcasassa@cecs.cl

L. Pablo Cid   Centro de Estudios Científicos (CECS), Avenida Arturo Prat 514, Casilla 1469, Valdivia, Chile, Tel 56 63 234500, Fax 56 63 234517, email pcid@cecs.cl

Claudio Colombano   Jefatura de Gabinete de Ministros, Buenos Aires, Argentina

Susana Díaz   Secretaría para la Tecnología y la Ciencia de la Presidencia de la Nación, Ushuaia, Tierra del Fuego, Argentina, Fax +54 2901 430644

Lydia Espizua   Instituto Argentino de Nivología, Glaciología y Estudios Ambientales (IANIGLA), Centro Regional de Investigaciones Científicas y Tecnológicas, Mendoza, Argentina, Fax +54 261 428 7370, email lespizua@lab.cricyt.edu.ar

María Angélica Godoi   Instituto de la Patagonia, Universidad de Magallanes, Casilla 113-D, Av. Bulnes 01855, Punta Arenas, Chile, Tel +56 61 207180, Fax +56 61 207187, email mangel@ona.fi.umag.cl

Niels Gundestrup   Department of Geophysics, University of Copenhagen, Juliane Maries Vej 30, DK-2100 Copenhagen, Denmark, Tel +45 35 32 05 53, Fax +45 35 36 53 57, email ng@gfy.ku.dk

Georg Kaser   Institut für Geographie, Universität Innsbruck, Innrain 52, A-6020 Innsbruck, Austria, Tel +43 512 507 5407, Fax +43 512 507 2895, email Georg.Kaser@uibk.ac.at

Shiro Kohshima   Laboratory of Biology, Faculty of Bioscience and Biotechnology, (% Fac. Science), Tokyo Institute of Technology, 2-12-1 O-okayama, Meguro-ku, Tokyo 152-8551, Japan, Tel +81 3 5734 2657, Fax +81 3 5734 2946, email kohshima@bio.titech.ac.jp

Pedro Labarca   Centro de Estudios Científicos (CECS), Avenida Arturo Prat 514, Casilla 1469, Valdivia, Chile, Tel +56 63 234500, Fax +56 63 234517

Ramón Latorre   Centro de Estudios Científicos (CECS), Avenida Arturo Prat 514, Casilla 1469, Valdivia, Chile, Tel +56 63 234500, Fax +56 63 234517

Juan Carlos Leiva   Instituto Argentino de Nivología, Glaciología y Estudios Ambientales (IANIGLA), Centro Regional de Investigaciones Científicas y Tecnológicas, Mendoza, Argentina, Fax +54 261 4285940, email jcleiva@lab.cricyt.edu.ar

Cristián Martínez   Centro de Estudios Científicos (CECS), Avenida Arturo Prat 514, Casilla 1469, Valdivia, Chile, Tel +56-63-234500, Fax +56-63-234515

Patricio Moreno   Departamento de Biología, Facultad de Ciencias, Universidad de Chile, Las Palmeras 3425, Ñuñoa, Santiago, Chile, Tel +56 2 678 7391, Fax +56 2 2712983, email pimoreno@uchile.cl

Hermann M. Niemeyer   Facultad de Ciencias, Universidad de Chile, Casilla 653, Santiago, Chile, Tel +56 2 678 7260, email niemeyer@abulafia.ciencias.uchile.cl

María Isabel Niemeyer   Centro de Estudios Científicos (CECS), Avenida Arturo Prat 514, Casilla 1469, Valdivia, Chile, Tel +56 63 234500, Fax +56 63 234517

Julio Poblete   Ministerio de Relaciones Exteriores, Santiago, Chile

Veijo Pohjola   Institutionen för Geovetenskaper (Department of Earth Sciences), Uppsala Universitet, Villavägen 16, 752 36 Uppsala, Sweden, Tel +46 18 471 2509, Fax +46 18 555 920, email veijo.pohjola@geo.uu.se

Margarita Préndez   Facultad de Ciencias Químicas y Farmacéuticas, Universidad de Chile, Casilla 233, Santiago, Chile, email mprendez@ll.ciq. uchile.cl

John N. Reeve   Department of Microbiology, The Ohio State University, Columbus, OH 43210-1292, USA, Tel +1 614 292 2301, Fax +1 614 292 8120, email reeve.2@osu.edu

Andrés Rivera   Centro de Estudios Científicos (CECS), and Departamento de Geografía, Universidad de Chile, P.O. Box 3387, Marcoleta 250, Santiago, Chile, Tel +56 2 6783032, Fax +56 2 2229522, email arivera@abello.dic.uchile.cl, arivera@cecs.cl

David Sabatini   Department of Cell Biology, New York University School of Medicine, 550 First Avenue, New York, NY 10016-6497, USA, Tel +1 212 263 5353, Fax +1 212 263 5813, email sabatd01@mcrcr.med.nyu.edu

Javier Sanz de Urquiza   Subsecretaría de Asuntos Latinoamericanos, Ministerio de Relaciones Exteriores, Comercio Internacional y Culto, Buenos Aires, Argentina, email jsu@mrecic.gov.ar

Margit Schwikowski   Paul Scherrer Institut, Labor für Radio- und Umweltchemie, CH-5232 Villigen PSI, Switzerland, Tel +41 56 310 4110, Fax +41 56 310 4435, email margit.schwikowski@psi.ch

Francisco V. Sepúlveda Centro de Estudios Científicos (CECS), Avenida Arturo Prat 514, Casilla 1469, Valdivia, Chile, Tel +56 63 234500, Fax +56 63 234517, email fsepulveda@cecs.cl

Rolf M. Sinclair 7508 Tarrytown Road, Chevy Chase, Maryland 20815-6027, USA, Tel +1 301 657-3441, email rolf@santafe.edu

Fernando C. Soncini Departamento de Microbiología, Facultad de Ciencias Bioquímicas y Farmacéuticas, Universidad Nacional de Rosario, Suipacha 531, 2000 - Rosario, Argentina, Tel +54 341 4370008, Fax +54 341 4804598, email patbact@citynet.net.ar

Claudio Teitelboim Centro de Estudios Científicos (CECS), Avenida Arturo Prat 514, Casilla 1469, Valdivia, Chile, Tel +56 63 234500, Fax +56 63 234515, email cecs@cecs.cl

Alejandro M. Viale Departamento de Microbiología, Facultad de Ciencias Bioquímicas y Farmacéuticas, Universidad Nacional de Rosario, Suipacha 531, 2000 - Rosario, Argentina, and, Instituto de Biología Molecular y Célular, CONICET, Argentina, Tel +54 341 4350661, email amviale@infovia.com.ar

Robert B. Waide LTER Network Office, Department of Biology, University of New Mexico, Albuquerque, NM 87131-1091, Tel +1 505 272 7311, Fax +1 505 272 7080, email rwaide@lternet.edu

Eske Willerslev Department of Evolutionary Biology, Zoological Institute, University of Copenhagen, Universitetsparken 15, 2100-DK Copenhagen, Denmark, Tel +45 35321309, Fax +45 35321300, email ewillerslev@zi.ku.dk

Mark Williams Institute of Arctic and Alpine Research, and Department of Geography, University of Colorado, Boulder, USA, email markw@snobear.colorado.edu

Participants on board the *Aquiles* for the Expedition-Conference in March 2000.

Panoramic view of southern front of Pío XI Glacier. Dark diagonal streak is a medial moraine with high volcanic content.

Oblique color satellite photograph of the Southern Patagonian Icefield, Chile, and Argentina. Photograph acquired on March 10, 1978, from Salyut-6 by Soviet cosmonauts G. M. Grechko and Yu. V. Romanenko, courtesy of Vladimir M. Kotlyakov. [Taken from "USGS Professional Paper 1386-I: Satellite Image Atlas of Glaciers of the World—South America," edited by Richard S. Williams, Jr., and Jane G. Ferrigno, pg. 1-165 (Fig. 31) (available at <http://pubs.usgs.gov/prof/p1386i/>)]

THE PATAGONIAN ICEFIELDS

GLACIER ECOSYSTEM AND BIOLOGICAL ICE-CORE ANALYSIS

Shiro Kohshima[1*], Yoshitaka Yoshimura[2], and Nozomu Takeuchi[3]

1. ABSTRACT

Biological activity on glaciers has been believed to be extremely limited. However, in Himalayan and Patagonian glaciers, we have found specialized biotic communities, including various cold-tolerant insects and copepods that were living on the glacier by feeding on algae and bacteria growing in the snow and ice. Since these microorganisms growing on the glacier surface are stored in the glacial strata every year, ice-core samples contain many layers with abundant microorganisms. In Himalayan glaciers, we studied snow algae on the glacier surface and in shallow ice cores, and showed that the layers with much snow algae in the ice core could be good boundary markers of annual layers and very useful for ice-core dating. Snow algae can be a new environmental proxy for studies of ice cores from warmer regions such as the Himalayas and Patagonia, where data on chemical and isotopic content data are not reliable because of heavy mixing.

2. INTRODUCTION

A glacier, a moving body of snow and ice, has long been believed to be an almost non-biological world because of its cold and severe environment, host only to temporary biotic communities based on wind-blown organic matter (Swan, 1961; Mani, 1962). However, persistent biotic communities with cold-tolerant animals and microorganisms especially adapted to the glacier environment, which complete their life cycles there, were discovered in Himalayan and Patagonian glaciers (Kohshima, 1984a,b; 1985a). A glacier is a relatively simple and closed ecosystem with a special biotic community comparable to those in other freshwater environments such as lakes and rivers.

[1] Laboratory of Biology, Faculty of Bioscience and Biotechnology (c/o Faculty of Science), Tokyo Institute of Technology, 2-12-1 O-okayama, Meguro-ku, Tokyo 152-8551, Japan. [2] Mitsubishi Kasei Institute of Life Sciences, 11 Minamiooya, Machida-shi, Tokyo 194-8511, Japan. [3] Frontier Observational Research System for Global Change, International Arctic Research Center, University of Alaska Fairbanks, 930 Koyukuk Dr. P.O.Box 757335 Fairbanks AK 99775-7335, U.S.A.

* corresponding author: kohshima@bio.titech.ac.jp

The Patagonian Icefields: A Unique Natural Laboratory for Environmental and Climate Change Studies. Edited by Gino Casassa et al., Kluwer Academic /Plenum Publishers, 2002.

Many ice-core studies have been done in various parts of the world to reconstruct past climate. However, it is only recently that the microorganism content of cores has come to be studied (Abyzov *et al.*, 1995; Willerslev *et al.*, 1999; Karl *et al.*, 1999; Priscu *et al.*, 1999), although many microorganisms, such as snow algae and bacteria, have been found on various glaciers in the world (Kol, 1942; Gerdel and Drouet, 1960; Kol and Flint, 1968; Kol, 1969; Kol and Peterson, 1976; Wharton *et al.*, 1981; Kohshima, 1984a,b; 1987a, 1989; Ling and Seppelt, 1993; Yoshimura *et al.*, 1997). There are very few reports on microorganisms that have lived on glaciers and have been preserved in glacier ice.

However, we found that the microorganisms growing on the Yala Glacier in the Langtang region of Nepal, were stored in the glacial strata every year and ice-core samples recovered from this glacier contained many layers, each possessing these microorganisms (Kohshima, 1984b; 1987a,b). Recently we studied a snow algal community on this glacier and showed that the biomass and the community structure of the snow algae clearly changed with altitude, reflecting the change of environmental conditions (Yoshimura *et al.*, 1997). We also analyzed the snow algae in a shallow ice core of this glacier and showed that they could be useful for ice-core analysis of this region (Yoshimura *et al.*, 2000).

This paper briefly reviews the results of these studies on glacier ecosystems and biological ice-core analysis.

3. GLACIER ECOSYSTEMS

3.1. Himalayan Glaciers

The first glacier found to have a permanent biotic community was the Yala (Dakpatsen) Glacier in the Langtang region of Nepal. The Yala Glacier is a plateau-shaped small glacier without a debris-covered area, with many flat terraced plateaus divided by crevasses and ice cliffs. The altitude of the glacier is between 5100 m and 5700 m. The equilibrium line of this glacier lies at an altitude of about 5300 m.

Two insect species (a chironomid and a springtail) and one species of crustacean (a copepod) were found on this glacier (Kohshima, 1984a; Kikuchi, 1994). The chironomid and the copepod live in the ablation area and the springtail lives in the accumulation area.

Adults of the newly found chironomid insect (*Diamesa kohshimai*), with reduced wings and antennae, are unable to fly (Figure 1). In the post-monsoon season, numerous adults were seen walking on the surface of the glacier when the sun was shining, but they quickly disappeared when it became overcast, moving freely down through the snow cover to the surface of the glacier ice. The larvae (Figure 1) were found living in the meltwater drainage channels running along the boundary of the glacier ice and the snow cover.

This species lives at remarkably low temperatures. Adult activity was observed on the surface during the day, when the snow surface temperature ranged from 0.0 °C to -7.2 °C. The temperature of the glacier ice surface 10 cm under the snow, where the adults stay at night, ranged from 0.0 °C to -12.8 °C. Furthermore, captured adults transported to a higher and colder place could walk slowly even at -16 °C. In contrast to their strong cold tolerance, they were very sensitive to warm temperatures. When placed on a human palm, they became hyperactive for a few seconds, but were paralyzed within about 20 seconds. When returned to the snow surface they recovered and began to walk again (Kohshima, 1984a).

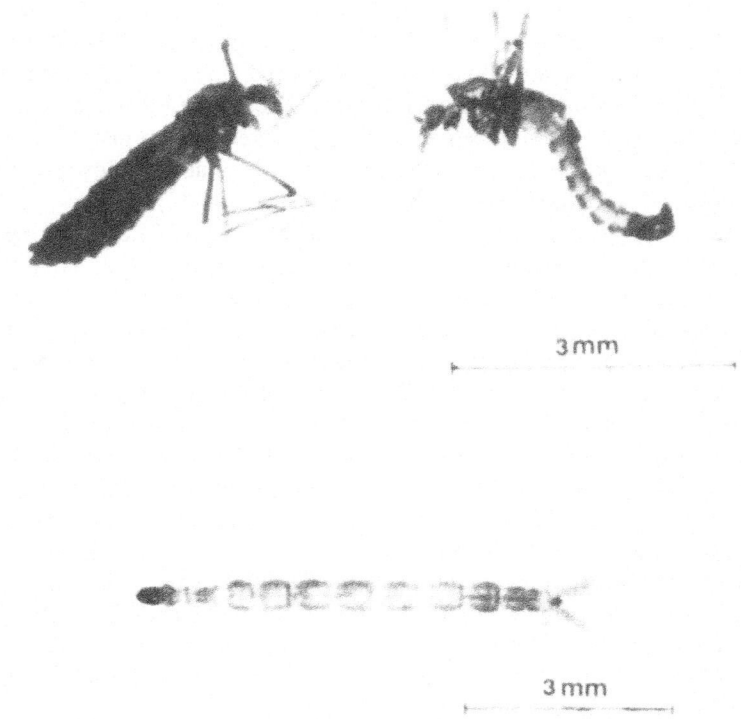

3mm

3mm

Figure 1. Female (top left) and male (top right) adults of the Himalayan glacier midge (*Diamesa kohshimai*). Larva (bottom) of the glacier midge.

Female adults of this insect were found to migrate toward the upper part of the glacier using a solar compass before laying eggs (Kohshima, 1985b). They walk straight using the solar compass and the walking direction could be altered by changing the apparent position of the sun with a hand mirror. Field data strongly suggest that the direction of their walk can be corrected by some information on the slope direction. They walked upstream, assessing the slope direction while walking straight by means of the solar compass. This seems to be an adaptive behavior against glacier movement and meltwater flow, to help glacier animals stay on the glacier.

All of the animals found on this glacier were feeding on algae and bacteria growing in the snow and ice. In this glacier, 11 species of snow algae were observed: *Chloromonas sp., Trochiscia sp., Mesotaenium berggrenii, Cylindrocystis brébissonii, Koliella sp., Ancylonema noldenskioeldii, Raphidonema sp.*, three *oscillatoriacean* algae and one unknown *coccoid* alga.

3.2. Patagonian Glaciers

A different type of glacial biotic community was found on Patagonian glaciers. Research was performed in outlet glaciers of the Northern (NPI) and Southern Patagonia Icefields (SPI) located at the southernmost tip of South America: the San Rafael Glacier (western side of the NPI), the Soler Glacier (eastern side of the NPI), the Nef Glacier (eastern side of the NPI) and the Tyndall Glacier (eastern side of the SPI). These glaciers are characterized by a temperate climate and fast glacial flow. At the terminus of the glaciers flowing into the sea, such as the San Rafael Glacier, the ice moves as fast as 16 m per day. In such places, the surface ice is heavily broken and no insects were found. However, at the higher points where glacier movement is relatively slow (about 1-2 m per day), many insects were found (Kohshima, 1985a).

One species of wingless stonefly (*Andiperla willinki;* Figure 2) and four species of springtail (*Isotoma spp.*) were found. On sunny days, many adult stoneflies and springtails were often observed actively moving about on the surface. Even when no insects were observed on the surface, many springtails were easily collected by breaking the surface ice. Many young wingless stonefly were found on the surface at night but very few were observed in the daytime. During the day, the young of this insect stayed in the deeper parts of water-filled crevasses or deep ice holes formed by meltwater (moulins). These structures are thought to be connected with large meltwater networks inside the glacier and these may be the main habitat of these organisms (Kohshima, 1985a). In the evening, the young came up to the surface and actively fed on springtails, algae, and bacteria. The adults also fed on springtails.

As the springtails feed on algae and bacteria, the food chain of this glacial biotic community is a little more complicated than that of the Yala Glacier. The surface ice of these glaciers contained abundant snow algae that occasionally cause "colored snow". However, the filamentous blue-green algae (*oscillatoriacean* algae), found on the Yala Glacier, were seldom seen in these Patagonian glaciers.

Figure 2. Adult of the Patagonian wingless stonefly (*Andiperla willinki*). The size of the insect is about 2 cm in length.

4. BIOLOGICAL ICE-CORE ANALYSIS

Snow algae in shallow ice cores from the Yala Glacier were examined for potential use in ice-core analyses. Shallow ice cores (7 m in length) taken at 5350 m asl of the Yala Glacier in 1994, contained more than 7 species of snow algae. In a vertical profile of the algal biomass, 11 distinct algal layers were observed (Figure 3).

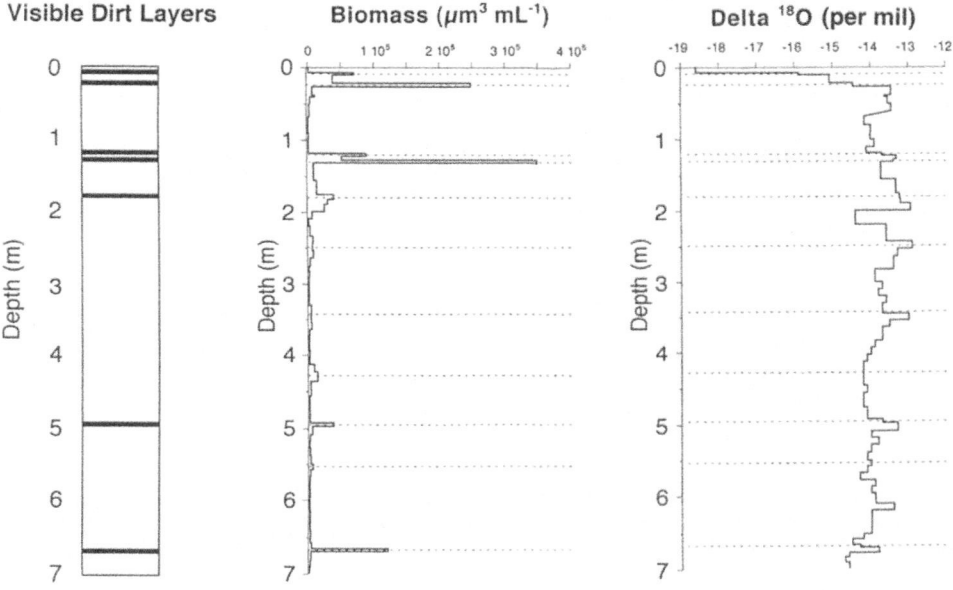

Figure 3. Vertical profile of the algal biomass in a shallow ice core (7 m in length) taken at 5350 m asl on the Yala Glacier in 1994. Eleven distinct algal layers were observed. Biomass is expressed in terms of unit volume, as $10^5 \ \mu m^3 \ mL^{-1}$.

Seasonal observations at the coring site in 1996, indicated that most algal growth occurred from late spring to late summer. Pit observation during 1991, 1992 and 1994, indicated algal layer formation was taking place annually. Delta ^{18}O, chemical ions (Na^+, Cl^-, SO_4^{2-}, and NO_3^-) and microparticles failed to show any clear seasonal variation, particularly so at depths exceeding 2 m, possibly due to heavy meltwater percolation (Figure 3). Snow algae in the ice cores would thus appear to be accurate boundary markers of annual layers and should prove very useful for ice-core dating in Himalayan-type glaciers (Yoshimura *et al.*, 2000).

The algal biomass on the Yala Glacier rapidly decreased as the altitude increased (Figure 4). The structure of the algal community represented by the proportion of each species to the total algal biomass also differed with altitude. From this difference in the algal community, we divided the glacier into three zones: the Lower Zone (5100-5200m asl; stable ice-environment) with seven species dominated by *Cylindrocystis brébissonii*, the Middle Zone (5200-5300 m asl; unstable transition area between ice-environment and snow-environment) with eleven species dominated by *Mesotaenium berggrenii,* and the Upper Zone (5300-5430 m asl; stable snow-environment) with four species dominated by *Trochiscia sp.* (Table 1).

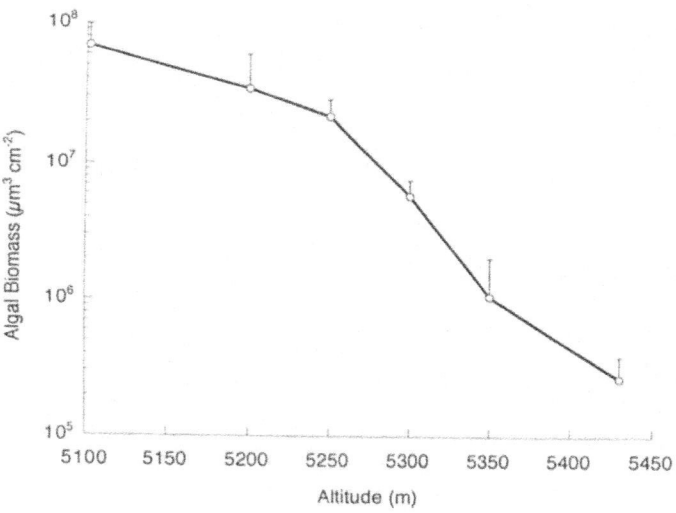

Figure 4. Change of the algal biomass with altitude observed in the Yala Glacier. Biomass rapidly decreased as the altitude increased. Biomass is expressed in terms of unit area.

Table 1. Characteristics of algal community and environment of each flora zone of Yala Glacier

Flora zone	Lower Zone	Middle Zone	Upper Zone
Altitude (m)	5100 - 5200	5200 - 5300	5300 - 5430
Dominant species	*Cylindrocystis brébissonii*	*Mesotaenium berggrenii*	*Trochiscia sp.*
Number of species	7	11	4
Species diversity	1.3 - 2.1	2.0 - 4.3	1.1 - 2.1
Environment	Ice-environment	Ice/Snow-environment	Snow-environment
Stability	Stable	Unstable	Stable
Substrate	Ice	Ice/Snow	Snow
Air temperature (°C)*	-0.1 - 2.4	-1.2 - 1.8	-2.1 - 0.9
Amount of water	Large	Middle	Small
Growth period	Long	Middle	Short
Snow cover (cm)	0	0 - 7.2	0 - 21.0
Light intensity	Strong	Strong/Weak	Weak
pH	4.9 - 5.4	4.9 - 5.5	5.2 - 5.9
Mineral particles (g m^{-2}: dry weight)	18 - 68	5.8 - 133	0.5 - 9.8

* Air temperature was estimated from meteorological data (Nepal Department of Hydrology and Meteorology, 1993) collected in Kyangjing (3920 m asl) and a lapse rate of 0.6 °C/100 m (after Yoshimura *et al.,* 1997).

This vertical zonation of the algal community type and biomass reflect the difference in summer climatic conditions with altitude. The relationship between algal biomass in the ice-core and meteorological data recorded near the glacier suggested that the algal biomass was well correlated with the mass balance during the summer, as estimated from air temperature and precipitation data. The mass balance during the summer is a proxy for the snow-layer thickness covering the snow algae. A thicker snow layer limits the radiation available for photosynthesis of the algae, thus reducing algal growth.

Thus, algal biomass and community structure in an ice core may reflect the summer conditions of past years. It is possible to detect past climate change by investigating changes in algal biomass and community structure in ice cores. The snow algae in the ice core can be a new information source for ice-core studies, especially in the analysis of ice-core samples from warmer regions such as Himalaya and Patagonia, where data on chemical and isotopic contents may not be reliable because of heavy mixing.

5. CONCLUSIONS

In one Himalayan glacier, three species of insect (two wingless chironomid and a springtail species) and one copepod, feed on aquatic blue-green algae and bacteria that grow on the glacier ice. In four outlet glaciers in the Patagonian icefields, the community consists of five species of insects (a wingless stonefly and four springtail species) and some microorganisms (snow algae and bacteria). All these cold-tolerant animals are especially adapted to the glacier environment and complete their life cycles in the glacier.

Snow algae in shallow ice cores from the Yala Glacier contained more than 7 species of snow algae, and 11 distinct layers with much snow algae were observed in a vertical profile of the algal biomass. The algal layers in the ice core were shown to be good boundary markers of annual layers and useful for ice-core dating. Since algal biomass and the type of algal community reflect the difference in summer climatic conditions on the glacier, the snow algae in the ice core can be a new information source for ice-core studies. This could be a good boundary marker of annual layers and signal of past climate change, which promises to be especially useful in the analysis of ice cores from warmer regions such as the Himalayas and Patagonia, where data on chemical and isotopic contents are not reliable because of heavy mixing related to melt.

6. REFERENCES

Abyzov, S. S., Barkov, M. I., Bobin, N. E., Lipenkov, V. Y., Mitskevich, I. N., Pashkevich, V. M., and Poglazova, M. N., 1995, Glaciological and microbiological description of the ice core in Central Antarctica, *Biology Bulletin*, **22**(5):441-446.

Gerdel, R.W., and Drouet, F., 1960, The cryoconite of the Thule area, Greenland, *Transactions of the American Microscopical Society*, **79**:256-272.

Karl, D. M., Bird, D. F., Bjokman, K., Houlihan, T., Shackelford, R., and Tupas, L, 1999, Microorganisms in the accreted ice of Lake Vostok, Antarctica, *Science*, **286**(5447):2144-2147.

Kikuchi, Y., 1994, *Glaciera*, a new genus of freshwater Canthocamptidae (Copepoda, Harpacticoida) from a glacier in Nepal, Himalayas, *Hydrobiologia*, **292/293**:59-66.

Kohshima, S., 1984a, A novel cold-tolerant insect found in a Himalayan glacier, *Nature*, **310**:225-227.

Kohshima, S., 1984b, Living micro-plants in the dirt layer dust of Yala Glacier, in: *Glacial Studies in Langtang Valley*, K. Higuchi, ed., Data Center for Glacier Research, Japanese Society of Snow and Ice, Nagoya University, pp.91-97.

Kohshima, S., 1985a, Patagonian glaciers as insect habitats, in: *Glaciological Studies in Patagonia Northern Icefield*, C. Nakajima, ed., Data Center for Glacier Research, Japanese Society of Snow and Ice, Nagoya University, pp. 96-99.

Kohshima, S., 1985b, Migration of the Himalayan wingless glacier midge (*Diamesa sp.*): Slope direction assessment by sun-compassed straight walk, *Journal of Ethology*, 3:93-104.

Kohshima, S., 1987a, Glacial biology and biotic communities, in: *Evolution and Coadaptation in Biotic Communities*, S. Kawano, J. H. Connell, and T. Hidaka, eds., Fac. of Science, Kyoto University, pp. 77-92.

Kohshima, S., 1987b, Formation of dirt layers and surface dust by micro-plant growth in Yala (Dakpatsen) Glacier, Nepal Himalayas, *Bulletin of Glacier Research*, 5:63-68.

Kohshima, S., 1989, Glaciological importance of microorganisms in the surface mud-like materials and dirt layer particles of the Chongce Ice Cap and Gozha Glacier, West Kunlun Mountains, China, *Bulletin of Glacier Research*, 7:59-65.

Kol, E., 1942, The snow and ice algae of Alaska, *Smithsonian Miscellaneous Collections*, 101:1-36.

Kol, E., 1969, The red snow of Greenland, II. *Acta Botanica Academiae Scientiarum Hungaricae*, 15(3-4):281-289.

Kol, E., and Flint, E. A., 1968, Algae in green ice from the Balleny islands, Antarctica, *New Zealand Journal of Botany*, 6:249-261.

Kol, E., and Peterson, J. A., 1976, Cryobiology, in: *The Equatorial Glaciers of New Guinea*, G. S. Hope, J. A. Peterson, U. Radok, and I Allison, eds., A. A. Balkema, Rotterdam, pp. 81-91.

Ling, H. U., and Seppelt, R. D., 1993, Snow algae of the Windmill Island, continental Antarctica, 2. *Chloromonas rubroleosa* sp. nov. (Volvocales, Chlorophyta), an alga of red snow, *Journal of Phycology*, 28:77-84.

Mani, M. S., 1962, *Introduction to High Altitude Entomology*, Methuen, London.

Priscu, J. C., Adams, E. E., Lyons, W. B., Voytek, M. A., Mogk, D. W., Brown, R. L., McKay, C. P., Takacs, C. D., Welch, K. A., Wolf, C. F., Kirshtein, J. D., and Avci, R., 1999, Geomicrobiology of subglacial ice above Lake Vostok, Antarctica. *Science*, 286(5447):2141-2144.

Swan, L. W., 1961, The ecology of the High Himalayas, *Scientific American*, 205:68-78.

Wharton, R. A. J., Vinyard, W. C., Parker, B. C., Simmons, G. M. J., and Seaburg, K. G., 1981, Algae in cryoconite holes on Canada Glacier in Southern Victorialand, Antarctica, *Phycologia*, 20(2):208-211.

Willerslev, E., Hansen, A. J., Christensen, B., Steffensen, J. P. and Arctander, P, 1999, Diversity of Holocene life forms in fossil glacier ice, *Proceedings of the National Academy of Sciences of the United States of America*, 96(14):8017-8021.

Yoshimura, Y., Kohshima, S., and Ohtani, S., 1997, A community of snow algae on a Himalayan glacier: change of algal biomass and community structure with altitude, *Arctic and Alpine Research*, 29(1):126-137.

Yoshimura, Y., Kohshima, S., Takeuchi, N., Seko, K. and Fujita, K., 2000, Himalayan ice core dating with snow algae, *Journal of Glaciology*, 46(153):335-340.

ISOLATION AND IDENTIFICATION OF BACTERIA FROM ANCIENT AND MODERN ICE CORES

Brent C. Christner[1], Ellen Mosley-Thompson[2], Lonnie G. Thompson[2], Victor Zagorodnov[2], and John N. Reeve[1*]

1. ABSTRACT

Glacial ice traps and preserves soluble chemical species, gases, and particulates including pollen grains, fungal spores and bacteria in chronologically-deposited archives. We have constructed an ice-core sampling system that melts ice only from the interior of cores, thereby avoiding surface contamination, and using this system we have isolated, cultured and characterized bacteria from ice cores that range from 5 to 20,000 years in age and that originate from both polar and non-polar regions. Low-latitude, high-altitude non-polar ice cores generally contain more culturable bacteria than polar ices, consistent with closer proximities to major biological ecosystems. Direct plating of melt-water from a 200-year old sample of ice from the Guliya ice cap on the Tibetan Plateau (China) generated ~180 bacterial colonies per ml [colony forming units/ml; (cfu/ml)], whereas meltwater from late Holocene ice from Taylor Dome in Antarctica contained only 10 cfu/ml, and <10 cfu/ml were present in ice of the same age from the Antarctic Peninsula and from Greenland. Based on their small-subunit ribosomal RNA-encoding DNA (rDNA) sequences many, but not all of the bacteria isolated are spore-forming species that belong to *Bacillus* and *Actinomycete* genera. Non-chronological fluctuations are observed in the numbers of bacteria present, consistent with episodic deposition resulting from attachment to larger particulates.

2. BACKGROUND

Snowfalls accumulate as glacial ice that traps gases, chemical species, and particulates in a chronological record that can be accessed by ice-core sampling. Ice cores collected

[1] Department of Microbiology, [2] Byrd Polar Research Center, Ohio State University, Columbus, OH, 43210-1292, USA.

* corresponding author: reeve.2@osu.edu

The Patagonian Icefields: A Unique Natural Laboratory for Environmental and Climate Change Studies
Edited by Gino Casassa et al., Kluwer Academic /Plenum Publishers, 2002.

from both polar and non-polar locations are archived at the Byrd Polar Research Center (BPRC) at the Ohio State University, and we are currently isolating bacteria from ice cores for which physical and chemical studies have already established ages and the climatic conditions prevailing when they were formed. The goals are to determine the numbers and types of bacteria recoverable from ice-cores from different locations, how these parameters changed in response to climate changes, and to obtain estimates of longevity of bacteria trapped in these frozen environments. The recovery of ice from the Guliya ice cap on the Tibetan Plateau (China) that is >500,000 years old (Thompson *et al.*, 1997), presents an opportunity to evaluate microbial survival in ice on a time scale that is meaningful for inter-planetary transport frozen within comets and also relevant to the idea that microbial life may continue to exist frozen below the surface of Mars or within the ice, or sub-ice ocean on Europa.

Previous reports have described the isolation of bacteria and fungi from glacial ice (Abyzov, 1993; Dancer *et al.*, 1997) and permafrost (Gilichinsky *et al.*, 1993; Shi *et al.*, 1997), with many isolates being species that differentiate naturally into radiation and desiccation-resistant spores or cysts. However, these are mostly reports of single sampling experiments whereas our goal is to compare the numbers and types of bacteria that can be recovered from ice of different ages from the same location, and from different ice cores of the same age from different global locations. Here we report the results of bacterial isolations from ice cores from Greenland, China, Bolivia and Antarctica, which range in age from 5 to 20,000 years.

3. SAMPLING TECHNOLOGY

The exterior surface of an ice core is inevitably contaminated during drilling and subsequent transport to the BPRC. An ice-core sampling system has therefore been developed and constructed to melt and collect melt-water samples only from the "non-handled" interior of an ice core (Figure 1). A slice is cut from the end of the core, and the cut surface is disinfected by soaking in 95% ethanol for 2 minutes at -20 °C, dissolving away an additional layer of ice. Control experiments in which bacteria were swabbed intentionally onto the cut surface and onto the saw blade used to cut the ice core, confirmed that such an ethanol wash effectively eliminated surface contamination. The core is then positioned vertically in the sampling device with the cut and disinfected end contacting the sampling head. By using different sized sampling heads, ice can be dissolved from one-half or one-quarter of the core facilitating repetition of the experiment, i.e., bacteria can be isolated from more than one sample of meltwater of the same age from the same ice core. The funnel-shaped sampling head is heated internally, the heat from the sampling head melts the contacted ice and the meltwater generated is collected through a central port at the base of the funnel. The heated sampling head moves vertically upwards through the interior of the ice core, generating and collecting meltwater only from the interior of the core. By using a fraction collector, meltwater from increasing depths within the core can be collected sequentially as separate samples.

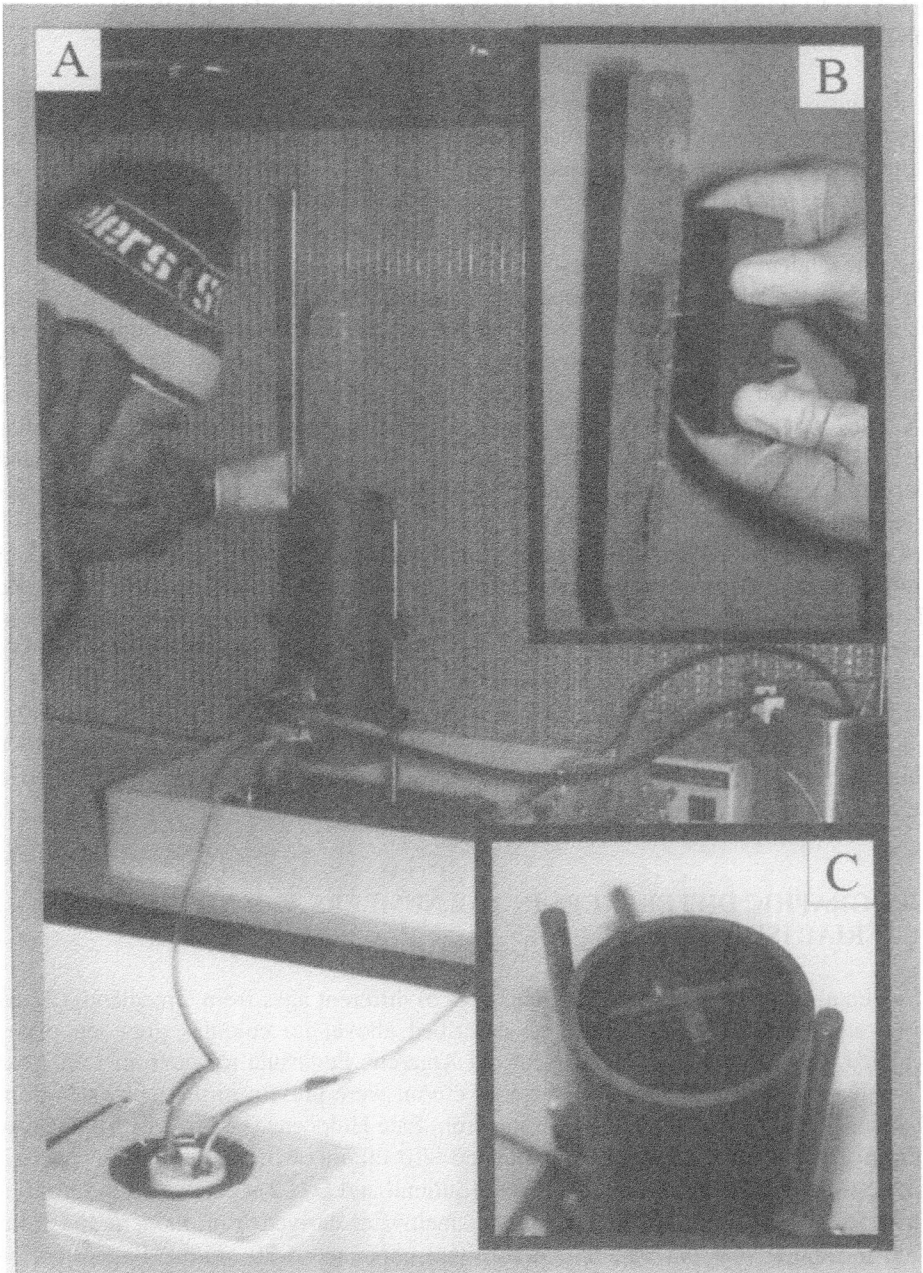

Figure 1. Ice-core sampler. A. The complete sampler with an ice core inserted in the ice-melting unit. All components of the system are sterilized by autoclaving and then assembled inside a laminar air-flow hood housed within a -20 °C walk-in freezer. B. The sampling head after movement through a core and removal of a cylindrical section from inside the core. C. Moveable dividers that facilitate melting through half or quarter sections of a core. Modified from Christner *et al.*, 2000. Reprinted with permission.

4. EVALUATION OF BACTERIAL GROWTH MEDIA AND GROWTH CONDITIONS

Plating samples of meltwater onto agar-solidified, commercial growth media containing low levels of nutrients, such as *Actinomycetes* isolation agar and R2A (Difco), or onto nutrient agar (Bacto) diluted 100-fold more than recommended by the manufacturer, resulted in the growth of more bacterial colonies than were obtained by plating on full-strength rich nutrient media. Such nutritionally-restricted conditions reduce the tendency of otherwise fast-growing species to dominate in bacterial isolations from environmental samples and apparently also remove an "immediate-growth" stress. This increases the survival and recovery of cells that have accumulated cellular damage during long periods of frozen inactivity. Some bacterial colonies appeared only after 1-2 months incubation at 25 °C, and some have appeared first after ~4 months incubation at 10 °C. Nevertheless, on subsequent sub-culture, most isolates form visible colonies after only 2-3 days incubation at 25 °C. All the isolates so far characterized have been isolated under aerobic growth conditions, but isolations under anaerobic growth conditions are currently in progress.

Initially each isolate is characterized phenotypically for colony color and morphology by light microscopy, growth temperature range, growth on different media and resistance/sensitivity to antibiotics. The 16S rDNA sequences corresponding to nucleotides 515 to 1492 of *Escherichia coli* 16S rDNA are then determined for all isolates exhibiting phenotypic differences. Based on such sequences, each isolate is identified in terms of its nearest, previously-characterized relative (Maidak *et al*, 1999; see Table 1). Although most phenotypically different isolates have different 16S rDNA sequences, some with different pigments that form colonies with different morphologies have been found to have identical 16S rDNA sequences.

5. GEOGRAPHIC DIFFERENCES IN THE NUMBERS AND DIVERSITY OF BACTERIAL ISOLATES

When aliquots of meltwater from ice cores of different ages from Greenland, China, Bolivia and Antarctica were plated as described above, no colonies grew on plates inoculated with meltwater from 150-year old Antarctic Peninsula ice or from 1500-year old Sajama ice (Bolivia), but ~180 bacterial cfu/ml were present in the meltwater from 200-year old Guliya ice (China). Meltwater from Late Holocene (~1,800 years old) polar ice from Taylor Dome (Antarctica) contained ~10 cfu/ml, whereas Late Holocene ice from the Antarctic Peninsula and Greenland (Summit and Dye 2) contained <2 cfu/ml. In contrast to the 1500-year old ice, aliquots of meltwater derived from modern and from 12,000 to 20,000-year old Sajama ice (Bolivia) contained 5-20 cfu/ml. Regardless of their geographic origin, many of the isolates formed highly pigmented colonies, consistent with pigments providing a survival advantage, presumably by increasing protection against solar irradiation during airborne distribution and exposure on the surface of a glacier.

Most of the isolates have 16S rDNA sequences closely related to non-sporulating Gram-positive bacteria, or to Gram-positive spore-forming *Bacillus* or *Actinomycetes* species. However, in total, bacteria that belong to *Acinetobacter, Arthrobacter,*

Aureobacterium, Bacillus, Brevibacterium, Cellulomonas, Clavibacter, Flavobacterium, Friedmanniella, Kurthia, Listeria, Microbacterium, Micrococcus, Micromonospora, Mycobacterium, Nocardioides, Paenibacillus, Propioniferax, Sphingomonas, Staphylococcus and *Stenotrophomonas* genera have been isolated based on their having 16S rDNA sequences that are ≥ 95% identical to that of previously characterized species (Table 1).

Table 1. Bacteria isolated from non-polar glacial ice (modified from Christner *et al.*, 2000; reprinted with permission).

Nearest Phylogenetic Neighbor[†]	Source and Age of Ice[‡]
Acinetobacter radioresistans	G50
Arthrobacter agilis	G5, G200
Arthrobacter globiformis	G200
Aureobacterium liquefaciens	G5
Aureobacterium testaceum	SB12K
Bacillus subtilis	G5, G200
Bacillus thuringiensis	G5, SB12K
Bacillus firmus	G200
Bacillus sporothermodurans	G200
Bacillus psychrophilis	SB100
Bacillus sp.10	G200
Brevibacterium acetylicum	SB12K
Cellulomonas turbata	G200
Cellulomonas hominis	SB12K
Clavibacter michiganensis	G5, G200, SB12K
Flavobacterium okeanokoites	G5
Friendmanniella antarctica	SB12K
Microbacterium aurum	G200
Microbacterium lacticum	SB12K
Micromonospora purpurea	G200
Micrococcus lylae	SB20K
Mycobacteria komossnese	SB12K
Norcardia corynebacteroides	SB12K
Norcardioides plantarum	SB12K
Paenibacillus amylolyticus	G5
Paenibacillus polymyxa	G5
Paenibacillus lautus	G200
Planococcus kocuri	SB150
Propioniferax innocua	G200
Staphylococcus aureus	G5
Stenotrophomonas africae	G5

[†] Based on 16S rDNA sequences being ≥ 95% identical to that of the named species, determined by the ShowDistance function of PAUP 4.0, beta version (Maidak *et al.*, 1999).
[‡] G50 and G200 designate 50 and 200-year old ice, respectively, from Guliya (China). SB100, SB150, SB12K and SB20K designate 100, 150, 12,000 and 20,000-year old ice, respectively, from Sajama (Bolivia).

Isolates from both 50 and 200-year old Guliya ice have 16S rDNA sequences that are > 99% identical to the 16S rDNA sequences of *Bacillus subtilis, Arthrobacter agilis* or *Clavibacter michiganensis* and, interestingly, an almost identical *C. michiganensis*-related isolate was obtained from 12,000-year old Sajama ice.

Plating meltwater also results in the growth of fungal colonies, consistent with the widespread airborne distribution of fungal spores, but attempts have not yet been made to identify the fungal species.

6. CONCLUSIONS

Bacteria revived from ice cores have probably endured desiccation and solar irradiation during airborne transport, followed by freezing, a period of frozen dormancy and thawing. We are currently determining whether our ice-core isolates have structural and/or biochemical features in common that can be related to their ability to survive such environmental abuse. Almost certainly, most such cells will have sustained some cellular damage, and the long incubation times required for initial colony formation are consistent with the idea these bacteria require a substantial period of time to repair accumulated cell damage before they can grow and divide successfully (Dodd *et al.*, 1997).

Most meltwater samples from ice cores from low-latitude, high-altitude glaciers in the Andes and Himalayas have been found to contain more culturable bacteria than samples from polar ices, consistent with these non-polar glaciers being closer to exposed soils and large biological ecosystems. Among the polar ices, the highest numbers of bacteria (~10 cfu/ml) have been recovered from Taylor Dome ice, from a site located at the head of the Taylor Valley in the Dry Valley complex of Antarctica. This is a very dry and very cold environment but there is substantial cryptoendolithic growth of algae, fungi and bacteria within the sandstone that dominates this region. Changes in the numbers of bacteria present within non-polar ice appear to be related to changes in climate. For example, in an earlier cooler, wetter period in South America, the abundance of local vegetation increased and presumably therefore the amounts of airborne particulates would have also increased. Particulates transport bacteria, and the result is an increased number of bacteria in Andean glacial ice formed at that time.

Most of the bacteria isolated to date are related to ubiquitous soil inhabitants, to species that have been isolated frequently in previous microbiological surveys of environmental samples from around the world. Many are species that differentiate into spores or cysts, cell types that tolerate environmental stress and survive extended periods of dormancy. For example, bacterial endospores remain viable for thousands of years (Abyzov, 1993), and possibly even for millions of years in amber (Cano *et al.*, 1995), although this has been disputed (Priest, 1995). Intriguingly, however, some of our isolates are closely related to species that were previously only isolated from frozen tundra soil, glacial ice, polar sea ice, or from the Antarctic Dry Valley region (Gosink and Staley, 1995; Bowman *et al.*, 1997; Schumann *et al.*, 1997; Shi *et al.*, 1997; Zhou *et al.*, 1997; Junge *et al.*, 1998; Siebert and Hirsch, 1988). The isolation of these species from geographically diverse but always cold and often frozen environments, suggests that they may have evolved cell structures and metabolic lifestyles specifically to grow and survive under such conditions. Identifying such cold-survival mechanisms would be inherently interesting, but would also provide important data for predicting the microbiology that might exist frozen in non-terrestrial environments.

7. ACKNOWLEDGMENTS

This research was supported by the National Science Foundation, Grant OPP-9714206.

8. REFERENCES

Abyzov, S. S., 1993, Microorganisms in the Antarctic ice, in: *Antarctic Microbiology*, E.I. Friedmann, ed., Wiley-Liss, New York, pp. 265-295.

Bowman, J. P., McCammon, S. A., Brown, M., V., Nichols D. S., and McMeekin, T. A., 1997, Diversity and association of psychrophilic bacteria in Antarctic sea ice, *Applied and Environmental Microbiology*, **63**:3068-3078.

Cano, R. J., and Boruki, M. K., 1995, Revival and identification of bacterial spores in 25- to 40-million-year-old Dominican amber, *Science*, **268**:1060-1064.

Christner, B. C., Mosley-Thompson, E., Thompson, L. G., Zagorodov, V., Sandman, K., and Reeve, J. N., 2000, Recovery and identification of viable microorganisms immured in glacial ice, *Icarus*, **144**:479-485.

Dancer, S. J., Shears, P., and Platt, D. J., 1997, Isolation and characterization of coliforms from glacial ice and water in Canada's High Arctic, *Journal of Applied Microbiology*, **82**:597-609.

Dodd, C. E. R., Sharman, R. L., Bloomfield, S. F., Booth, I. R., and Stewart, G. S. A. B., 1997, Inimical processes: bacterial self-destruction and sub-lethal injury, *Trends in Food Science and Technology*. **8**:238-241.

Gilichinsky, D. A., Soina, V. S., and Petrova, M. A., 1993, Cryoprotective properties of water in the Earth cryolithosphere and its role in exobiology, *Origin Life Evolution Biosphere*, **23**:65-75.

Gosink, J. J., and Staley, J. T., 1995, Biodiversity of gas vacuolate bacteria from Antarctic sea ice and water, *Applied and Environmental Microbiology*, **61**:3486-3489.

Junge, K., Gosink, J. J., Hoppe, H.-G., and Staley, J. T., 1998, *Arthrobacter, Brachybacterium* and *Planococcus* isolates identified from Antarctic sea ice brine. Description of *Planococcus mcceekinii*, sp. nov. , **21**:306-314.

Maidak, B. L., Cole, J. R., Parker Jr., C. T., Garrity, G. M., Larsen, N., Li, B., Lilburn, T. G., McCaughey, M. J., Olsen, G. J., Overbeek, R., Pramanik, S., Schmidt, T. M., Tiedje, J. M., and Woese, C. R., 1999, A new version of the RDP (Ribosomal Database Project). *Nucleic Acids Research*, **27**:171-173.

Priest, F. G., 1995, Age of bacteria in amber, *Science*, **270**:2015-2016.

Schumann, P., Prauser, H., Rainey, F. A., Stackebrandt, E., and Hirsch, P., 1997, *Friedmanniella antarctica* gen. Nov., sp. nov., and LL-diaminopilelic acid-containing actinomycete from antarctic sandstone, *International Journal of Systematic Bacteriology*, **47**:278-283.

Shi, T., Reeves, R. H., Gilichinsky, D. A., and Friedmann, E. I., 1997, Characterization of viable bacteria from Siberian permafrost by 16S rDNA sequencing, *Microbiological Ecology*, **33**:169-179.

Siebert, J., and Hirsch, P., 1988, Characterization of 15 selected coccal bacteria isolated from antarctic rock and soil samples from the McMurdo - Dry Valleys (South Victoria Land), *Polar Biology*, **9**:37-44.

Thompson, L. G., Yao, T., Davis, M. E., Henderson, K. A., Mosley-Thompson, E., Lin, P. N., Beer, J., Synal, H. A., Cole-Dai, J., and Bolzan, J. F., 1997, Tropical climate instability: the last glacial cycle from the Qinghai-Tibetan Plateau, *Science*, **276**:1821-1825.

Zhou, J., Davey, M. E., Figueras, J. B., Rivkina, E., Gilichinsky, D., and Tiedje, J. M., 1997, Phylogenetic diversity of a bacterial community determined from Siberian tundra soil DNA, *Microbiology*, **143**:3913-3919.

PERSPECTIVES FOR DNA STUDIES ON POLAR ICE CORES

Anders J. Hansen[1†] and Eske Willerslev[1†*]

1. ABSTRACT

Recently amplifiable ancient DNA was obtained from a Greenland ice core. The DNA revealed a diversity of fungi, plants, algae and protists and has thereby expanded the range of detectable organic material in fossil glacier ice. The results suggest that ancient DNA can be obtained from other ice cores as well. Here, we present some future perspectives for DNA studies on polar ice cores in regard to molecular ecology, DNA damage and degradation, anabiosis and antibiotic resistance genes. Finally, we address some of the methodological problems connected to ancient DNA research.

2. INTRODUCTION

In recent decades, numerous ice cores have been drilled in many parts of the world including Greenland, Canada, Russia and the Antarctic (Figure 1a, b). The ice cores range from a few hundred meters to more than three kilometers in length and constitute a record of continual snow-accumulation as far back as 400,000 years B.P.

The main purpose of ice-core drilling has been the study of paleoatmospheric chemistry yielding detailed information about climatic changes in the past (Dansgaard *et al.*, 1993). However, organic material has also been found in several of the ice cores (Hammer *et al.*, 1985, Kumai and Langway, 1988), but only plant pollen grains and spores have been characterized (Fredskild and Wagner, 1974; MacAndrews *et al.*, 1984; Bourgeois, 1986). Most of the organic material in the ice does not lend itself to morphological identification and therefore, little is known of the diversity of organisms in fossil glacier ice (Figure 2a, b).

---

[1] University of Copenhagen, Department of Evolutionary Biology, Universitetsparken 15, DK-2100 Copenhagen, Denmark

[†] The authors have contributed equally to the work and should be regarded as joint first authors.

[*] corresponding author: ewillerslev@zi.ku.dk

The Patagonian Icefields: A Unique Natural Laboratory for Environmental and Climate Change Studies
Edited by Gino Casassa et al., Kluwer Academic /Plenum Publishers, 2002.

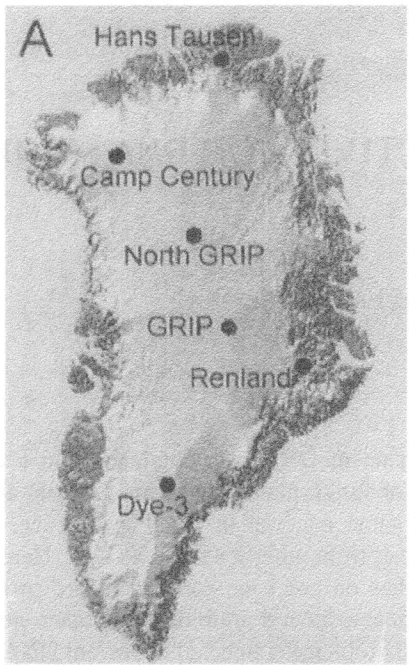

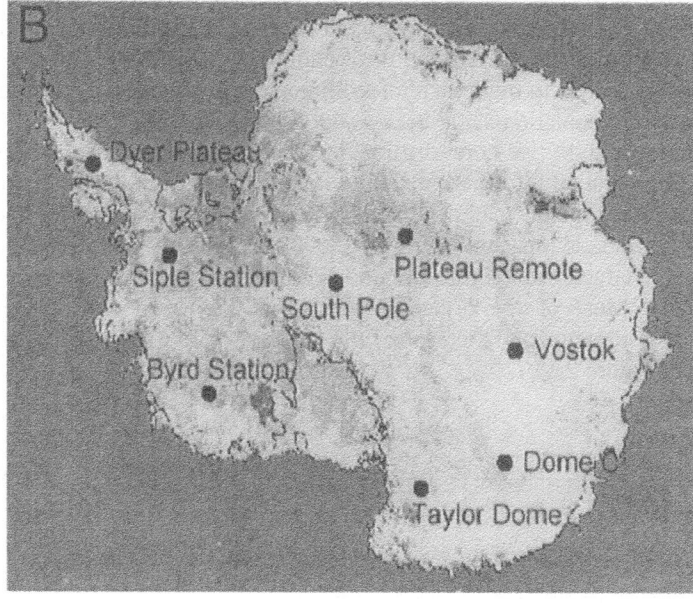

Figure 1. The location of some ice-core drilling sites in **A,** Greenland and **B,** Antarctica (Modified from the Geodetic Institute of Copenhagen 1963 and http://www-bprc.mps.ohio-state.edu/Icecore/Antarctica.html).

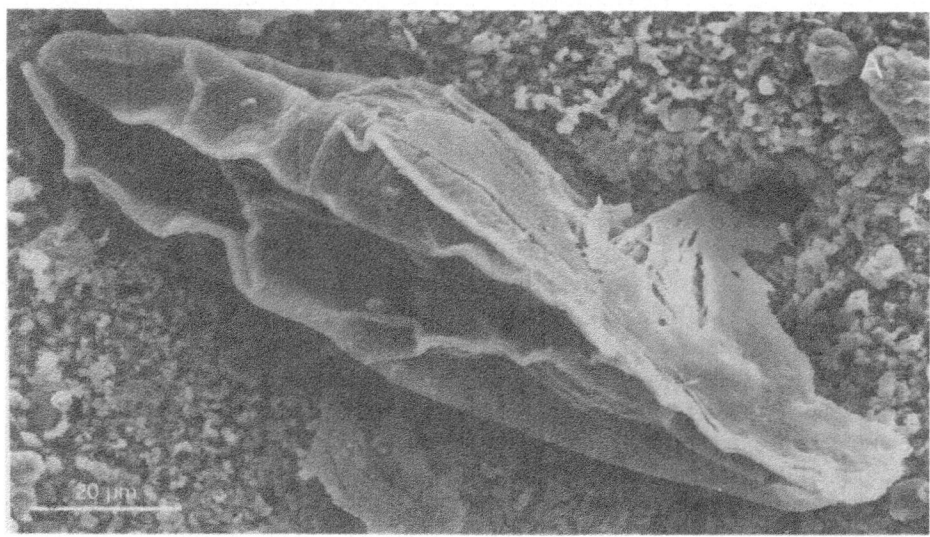

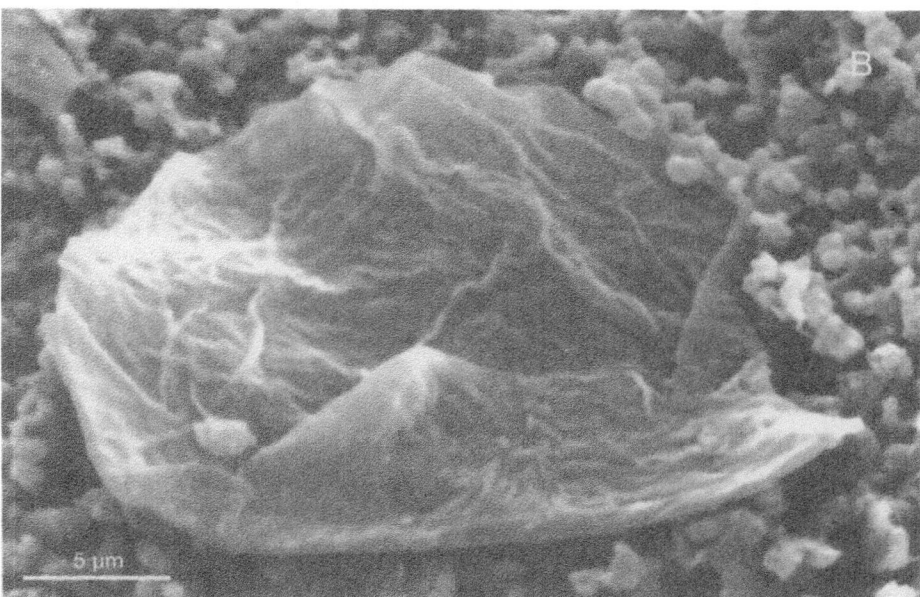

Figure 2. Scanning Electron Microscope (SEM) photos of unidentifiable material (organic?) from 4000-year old glacier ice from the Hans Tausen ice core. The material was concentrated by filtration. The filter paper was transferred to an aluminium stub and sputtercoated with gold. All SEM examinations were performed with a JEOL-840 microscope.

The invention of the Polymerase Chain Reaction (PCR) (Saiki *et al.,* 1985; Mullis and Faloona, 1987) has made it possible to obtain DNA from a large number of hitherto unknown organisms enabling characterization of several poorly described communities (Giovannoni *et al.,* 1990; Ward *et al.,* 1990; Liesack and Stackebrandt, 1992). As neutral pH and low temperatures that prevail in ice cores are believed to be critical for long-term preservation of DNA (Pääbo *et al.,* 1989; Höss *et al.,* 1996; Cooper *et al.,* 1997), we recently attempted to study the composition of eukaryotic organisms in fossil glacier ice by the means of PCR-based molecular techniques.

From 2000- and 4000-year-old ice-core samples drilled at the Hans Tausen ice cap in the northernmost part of Greenland (82.5° N, 37.5° W, 1270 m above sea level), we obtained and characterized 120 clones that represent at least 57 distinct taxa and revealed a diversity of fungi, plants, algae and protists of both local and distant origin (Willerslev *et al.,* 1999; Table 1). The study, thereby, expanded the range of detectable organic material in fossil glacier ice, and the results suggest that ancient DNA can be obtained from other ice cores as well (Frank, 1999). Therefore, molecular investigations could provide a platform for future studies of deep ice cores and possibly yield valuable information about ancient communities and their changes over time.

From the few results currently obtained it is difficult to delimit the possibilities for molecular studies on ice cores. In any case, in light of the purpose of our meeting, we find it reasonable to present what we believe could be future possibilities for DNA investigations on ice cores.

Table 1. Clone sequences from the Hans Tausen ice core identified to the taxonomic level of class, order or family (Modified from Willerslev *et al.,* 1999).

Kingdom	Phyla	Class	Order or family
Fungi	*Ascomycota*	*Euascomycetes*	*Ophiostomataceae*
			Eurotiales
			Sordariales
			Dorthideales
		Hemiascomycetes	*Saccharomycetaceae*
			Chaetothyriales
	Basidiomycota	*Urediniomycetes*	*Sporidiales*
		Hymenomycetes	*Hymenochateraceae*
Viridiplantae	*Cholorophyta*	*Chlorophyceae*	*Chlamydomonadaceae*
	Coniferophyta	*Coniferopsida*	*Pinaceae*
"Alveolata"	*Ciliophyta*	*Oligohymenophora*	-
"Stramenopiles"	-	*Chrysophyceae*	-

3. MICROBIAL DIVERSITY OVER TIME

With a thorough method for extraction and analysis of ancient DNA from fossil glacier ice and access to ice-core samples from both the northern and southern hemisphere, it has now become possible to make comparative studies of microbial diversity, past and present. A possible approach for future DNA studies on ice cores could therefore include:

i. Investigation of the diversity of microorganisms (prokaryotes and microbial eukaryotes), past and present, through comparative studies of DNA in ice cores drilled at different geographical locations.
ii. Investigation of the possible relationships of microbial diversity (functional and taxonomic) to climatic parameters, such as temperature, dust content, precipitation, etc., which are already available from ice-core studies.

In this way, it could be possible to retrieve information concerning climate change and its effect on microbial diversity over time. This type of investigation could also be of interest to, for example, the debate on global change and its effect on biological communities.

The analysis would require several distinct samples from several ice cores representing different eras. Other markers in addition to the 18S rRNA gene previously used (Willerslev *et al.,* 1999), could be the Internal Transcribed Spacer (ITS) regions (ITS1, ITS2) and 5.8S rDNA for identification of microbial eukaryotes, and the large subunit of the ribulose-1.5-biphosphate carboxylase (*rbc*L) for identification of the remains of higher plants. These regions are well represented in the sequence databases and evolve faster than the 18S rRNA gene, thereby, making it possible to identify clone sequences to lower taxonomic levels (Rollo *et al.,* 1994). The 16S rRNA gene could be used to study prokaryote diversity. Correlations between the obtained diversity patterns and climatic data could then be performed using statistical methods such as regression analysis (Willerslev *et al.,* in press) in an attempt to isolate possible variables responsible for the diversity patterns observed.

Other diversity studies on ice cores could include specific hypotheses related to fluctuations in atmospheric methane and methanesulphonate.

4. DIVERSITY OF METHANOGENS

Greenland ice cores, as well as those from Antarctica, indicate that large changes in the atmospheric concentration of greenhouse gases, such as carbon dioxide and methane, have occurred during the last 160,000 years. The studies further suggest that glacial and interglacial periods are associated with low and high methane values, respectively. The reason for the fluctuations in atmospheric methane is unknown but may be explained by an expansion and reduction of tropical wetlands, a habitat for methane-producing bacteria (methanogens; Chappellaz *et al.,* 1990; Brook *et al.,* 1996). PCR-based molecular techniques now make it possible to investigate:

i. Whether there are any methanogens present in ice cores from the northern and southern hemispheres.
ii. Whether variations in methanogen diversity can be correlated to the observed fluctuations in atmospheric methane.

5. DIVERSITY OF DMS-PRODUCING PHYTOPLANKTON

Dimethylsulfide (DMS) creates sulfuric acid and methanesulfonate (MSA) in the atmosphere and is likely to have influence on global climate regulation (Andreae, 1990). DMS-producing marine phytoplankton, such as *Coccolithiphora*, are believed to be the most prominent biogenic source of DMS (Keller *et al.*, 1989). Ice-core studies have shown that the atmospheric concentration of MSA has fluctuated during the last 100,000 years. While ice cores from Antarctica show the highest levels of DMS during cold periods, the opposite trend is found in the Greenland ice cores (Legrand *et al.*, 1991; Saltzman *et al.*, 1997). The reasons for the fluctuations in atmospheric DMS, as well as the clear difference between the two polar regions, are unknown. However, a common suggestion is that changes in the distribution and composition of DMS-producing marine phytoplankton over time are responsible. Now, molecular techniques make it possible to investigate:

i.   Whether DMS-producing phytoplankton are present in ice cores from Greenland and Antarctica.

ii. Whether changes in the number and composition of DMS-producing phytoplankton over time can be correlated to the observed fluctuations in atmospheric MSA.

6. BASE-MISINCORPORATIONS

Modifications in ancient DNA have previously been identified and quantified by gas chromatography/mass spectrometry (GC/MS) (Höss *et al.*, 1996). As several ice cores exceed DNA's theoretical life span of approximately 100,000 years (Poiner *et al.*, 1996; Krings *et al.*, 1997) and cold burial conditions are believed to be critical for long-term preservation of DNA (Höss *et al.*, 1996; Cooper *et al.*, 1997), ice cores would probably be suitable media for studying DNA degradation over time. However, GC/MS requires relatively large amounts of DNA ($\geq 1\mu g$), which are unlikely to be present in ice-core samples (Pääbo pers. comm.). Nevertheless, damage from, for example, hydrolytic and oxidative miscoding lesions, leads to misincorporations of nucleotides during amplification and these will therefore be evident in sequences from cloned PCR products (Greenwood *et al.*, 1999). As more sequence material becomes available from diversity studies on ice cores, it will become possible to statistically analyze miscoding lesions in DNA from fossil glacier ice as has already been performed on DNA from ancient human remains (Hansen *et al.*, 2001). However, to avoid interference from possible undamaged DNA belonging to viable microbes that may be present in the samples (see below), one should focus on DNA from fungi that have been cultured from ice-core samples less that 11,000 years old (Abyzov *et al.*, 1982). An investigation of base-misincorporations in DNA from fossil glacier ice is likely to yield new information relevant to the discussion of DNA damage and degradation for the following reasons:

i.   Low temperatures and neutral pH that prevail in ice cores are believed to be critical for long-term preservation of DNA (Pääbo *et al.*, 1989; Höss *et al.*, 1996; Cooper *et al.*, 1997).

ii. Ice cores constitute a unique and important historical record as organic material is deposited in almost identical conditions throughout time.

iii. Several ice cores are of an age that exceeds a time-span of more than 100,000 years, believed by many to be the maximum life-span of amplifiable ancient DNA (Poiner *et al.*, 1996; Krings *et al.*, 1997).

7. AMINO ACID RACEMIZATION

All amino acids used in proteins, with the exception of glycine, can exist in the form of two optical isomers, the D- and L-enantiomers, of which the L-form is used exclusively in protein biosynthesis. Once isolated from active metabolic processes, the L-form undergoes racemization to produce D-amino acids until, eventually, the L- and the D-form of a particular amino acid are present in equal amounts (Bada, 1972). Amino acid racemization is affected by some of the same factors that affect spontaneous hydrolysis of DNA (Bada, 1972; Lindahl and Nyberg, 1972; Lindahl, 1993). A rough correlation between the extent of amino acid racemization and the length of retrievable DNA sequences from fossil remains has been reported, but no correlation has been observed between either of these factors and sample age (Poiner *et al.*, 1996), probably due to differences in preservation conditions of the specimens investigated.

In general, analysis of amino acid racemization using high-pressure liquid chromatography (HPLC) requires only a few milligram of ancient material (Poiner *et al.*, 1996). For this reason, it is feasible that amino acids from ice cores could be investigated using this type of analysis. As the amino acids and the DNA in ice caps are preserved through time under nearly homogeneous conditions, ice cores could be suitable for:

i. Studying correlations between amino acid racemization and age, frequency of base-misincorporations (see above) or length of amplification products.

ii. Investigating possible correlations between age and frequency of base-misincorporations or fragment length.

8. ANABIOSIS

There have been several recent reports of anabiosis, i.e., restoration of life from a seemingly dead condition, of ancient microorganisms (Kennedy *et al.*, 1994; Cano and Borucki, 1995; Stone 1999). Some of these spectacular findings concerned the culturing of bacteria and yeast-like fungi from ice-core samples of over 100,000 years old (Abyzov, 1993; Christner *et al.*, 2000). These organisms are believed to have survived the pressure of the ice by associating themselves with protective particles, such as pollen grains and tiny dirt pockets. However, like other reports of anabiosis, these have been met with scepticism (Eglinton, 1996) because such reports depend solely on circumstantial evidence (Ubaldi *et al.*, 1996).

It has become possible to use DNA analysis to shed light on the possibility of retrieving viable microorganisms from fossil glacier ice. We contend that if a certain type of cell survived prolonged freezing as it is claimed, it would be possible to find the remains of cells of the same type, which did not survive such conditions. In other words, each living cell should be accompanied by a relatively large number of mummified cells.

Thus, we may propose an experimental investigation to study whether DNA from variable microorganisms cultured from a given ice-core sample, match some of the DNA retrieved in a separate laboratory from the same samples. Positive results from such an experiment would strongly support the hypothesis of long-term anabiosis in ice caps.

9. ANTIBIOTIC-RESISTANT GENES

The widespread use of antibiotics in human medicine and in animal husbandry during recent decades has resulted in a rapid dissemination of antibiotic-resistant genes amongst bacterial populations (Speer *et al.*, 1992). Although little is known about the mechanisms behind this rapid spread of antibiotic resistance, two main hypotheses have been proposed:

  i.   Antibiotic-resistant genes in clinical isolates originate from naturally occurring antibiotic-producing microorganisms, which need the mechanism for self-protection, and have spread by various mechanisms of lateral gene transfer (van Elsas, 1992).
  ii. Alternatively, antibiotic-resistant genes have evolved from bacterial genes encoding enzymes involved in normal cellular metabolism (Shaw *et al.*, 1993).

Ice cores make possible the study of bacterial genes before and after the widespread use of antibiotics in human and animal care. On the basis of well-characterized tetracycline-resistant genes, one could investigate:

  i.   Whether any fragments of known tetracycline-resistant genes are present in the ice-core samples.
  ii. Whether there is any difference in the diversity of tetracycline resistance genes before and after the medical use of tetracycline.

Together with an identification of the bacteria in question, such a study is likely to throw new light on the evolution of the tetracycline-resistant genes.

10. ICE-CORE GENETICS – A METHODOLOGICAL CHALLENGE

Amplification of ancient DNA is beset with difficulties, which can be broadly categorized into three areas:

  i.   Modified old DNA templates as well as other components in extracts of old tissues are likely to inhibit the DNA polymerase enzyme leading to false negative results (Pääbo, 1990).
  ii. DNA template damage can cause amplification artefacts such as base-misincorporations (see above) and the generation of chimera sequences by "jumping PCR" events (Pääbo *et al.*, 1989). This, together with the fact that fossil remains do not allow amplification of endogenous DNA that has more than a few hundred base pairs, makes it difficult to interpret an ancient DNA sequence in biological terms (Handt *et al.*, 1994a, Hagelberg *et al.*, 1994).
  iii. The great sensitivity of the amplification process and the small quantities of starting templates in fossil remains makes the PCR liable to pick up even minor

contaminations of undamaged contemporary DNA, which are then preferentially amplified leading to false positive results (Austin *et al.*, 1997). The contaminating DNA can enter the PCR process at any stage of sample processing, for example, within the samples themselves, chemical reagents, laboratory disposables, or from the surrounding air (Schmidt *et al.*, 1995). This problem has forced workers to establish cumbersome procedures that require numerous precautions and controls, such as physical separation of pre- and post-PCR laboratories and strict adherence to several criteria, such as the requirement for independent replication of results, before a DNA sequence determined from extracts of an ancient specimen can be considered as genuine (Pääbo *et al.*, 1989; Pääbo, 1990; Handt *et al.*, 1994b, Cooper, 2001).

11. CONCLUSION

Ancient DNA studies on ice cores are still in their infancy and it is therefore difficult to predict their future possibilities and limitations. It may be possible to address several topics within microbial ecology, DNA damage and degradation, anabiosis and antibiotic resistance genes using studies of ice-core genetics. However, the risk of contamination with contemporary DNA necessitates numerous precautions and fulfillment of strict criteria, making laboratory work time consuming and costly.

12. ACKNOWLEDGMENTS

We thank I. Barnes, J. Bourgeois, B. Christensen, A. Cooper, N. Daugbjerg, D. Fisher, R. Koener, S. Mathiasen, S. Pääbo, T. Quin, J. P. Steffensen, S. Sørensen, C. Wiuf and C. M. Zdanowicz for help and discussions; and The Velux Foundation of 1981, Denmark for financial support.

13. REFERENCES

Abyzov, S. S., 1993, Microorganisms in the Antarctic ice, in: *Antarctic Microbiology*, E. I. Friedmann, ed., Wiley-Liss, Inc. USA, pp. 265-295.

Abyzov, S. S., Bobin, N. E., and Kudryashov, B. B., 1982, Quantitative analysis of microorganisms during the microbial investigation of Antarctic glaciers, *Izvestiya Akademii Nauk SSSR, Seriya Biologicheskaya*, 6:897-905. (in Russian).

Andreae, M. O., 1990, Ocean-atmosphere interactions in the global biogeochemical sulphur cycle, *Marine Chemistry*, 30:1-29.

Austin, J. J., Ross, A. J., Smith, A. B., Fortey, R. A., and Thomas, R. H., 1997, Problems of reproducibility: Does geologically ancient DNA survive in amber-preserved insects? *Proceedings of the Royal Society London, Series B.*, 264:467-474.

Bada, J. L., 1972, Kinetics of racemization of amino acids as a function of pH, *Journal of the American Chemical Society*, 94:1371-1373.

Bourgeois, J. C., 1986, A pollen record from the Agassiz Ice Cap, northern Ellesmere Island, Canada, *Boreas*, 15:345-354.

Brook, E. J., Sowers, T., and Orchardo, J., 1996, Rapid variations in atmospheric methane concentrations during the past 110,000 years, *Science*, 273:1087-1090.

Cano, R. J., and Borucki, M. K., 1995, Revival and identification of bacterial spores in 25- to 40-million-year-old Dominican amber, *Science*, 268:1060-1064.

Chappellaz, J., Barnola, J. M., Raynaud, D., Korotkevich, Y. S., and Lorius, C., 1990, Ice-core record of atmospheric methane over the past 160,000 years, *Nature,* **345**:127-131.

Christner, B. C., Mosley-Thompson, E., Thompson, L. G., Zagorodov, V., Sandman, K., and Reeve, J. N., 2000, Recovery and identification of viable microorganisms immured in glacial ice, *Icarus,* **144**:479-485.

Cooper, A., Poiner, H. N., Pääbo, S., Radovic, J., Debenath, A., Caparros, M., Barroso-Ruiz, C., Bertranpetit, J., Nielsen-Marsh, C., Hedges, R. E., and Sykes, B., 1997, Neanderthal genetics, *Science,* **277**:1021-1024.

Cooper, A., Rambaut, A., Macaulay, V., Willerslev, E., Hansen, A. J., and Stringer, C., 2001, Human origins and ancient human DNA, *Science,* **292**:1655-1656.

Dansgaard, W., Johnsen, S. J., Clausen, H. B., Dahl-Jensen, D., Gundestrup, N. S., Hammer, C. U., Hvidberg, C. S., Steffensen, J. P., Sveinbjörnsdóttir, A. E., Jouzel, J., and Bond, G. C., 1993, Evidence for general instability of past climate from a 250-kyr ice-core record, *Nature,* **364**:218-220.

Eglinton, G., 1996, Some thoughts on: Life detection on Earth and the limits of microbial life, *ABI Newsletter,* 3:10.

Frank, L., 1999, Cool DNA, *Science,* **285**:327.

Fredskild, B., and Wagner, O., 1974, Pollen and fragments of plant tissue in core samples from the Greenland Ice Cap, *Boreas,* 3:105-108.

Giovannoni, S., Britschgi, T., and Feld, K., 1990, Genetic diversity in Sargasso Sea bacterioplankton, *Nature,* **345**:60-63.

Greenwood, A. D., Capelli, C., Possnert, G., and Pääbo, S., 1999, Nuclear DNA sequences from late Pleistocene megafauna, *Molecular Biology and Evolution,* **16**:1466-1473.

Hagelberg, E., Thomas, M. G., Cook, C. E. Jr, Sher, A.V., Baryshnikov, G. F., and Lister, A. M., 1994, DNA from ancient mammoth bones, *Nature,* **370**:333-334.

Hammer, C. U., Clausen, H. B., Dansgaard, W., Neftel, A., Kristinsdottir, P., and Johnson, E., 1985, Continuous impurity analysis along the Dye 3 deep core, *Geophysical Monograph,* **33**:90-94.

Handt, O., Richards, M., Trommsdorff, M., Kilger, C., Simanainen, J., Georgiev, O., Bauer, K., Stone, A., Hedges, R., and Schaffner, W. 1994a, Molecular genetic analyses of the Tyrolean Ice Man, *Science,* **264**:1775-1778.

Handt, O., Höss, M., Krings, M., and Pääbo, S., 1994b, Ancient DNA: methodological challenges, *Experientia,* **50**:524-529.

Hansen, A. J., Willerslev, E., Wiuf, C., Mourier, T., and Arctander, P., 2001, Statistical evidence miscoding lesions in ancient DNA templates, *Molecular Biology and Evolution,* **18**:262-265.

Höss, M., Jaruga, P., Zastawny, T. H., Dizdaroglu, M., and Pääbo, S., 1996, DNA damage and DNA sequence retrieval from ancient tissues, *Nucleic Acids Research,* **24**:1304-1307.

Keller, M. D., Bellows, W. K., and Guillard, R. R. L., 1989, Dimethyl sulfide production in marine phytoplankton, in: *Biogenetic sulphur in the environment,* E. Saltzman and W. J. Cooper, eds., American Chemical Society Symposium Series, Washingon DC, **393**:167-182.

Kennedy, M. J., Reader, S. L., and Swierczynski, L. M., 1994, Preservation records of micro-organisms: evidence of the tenacity of life, *Microbiology,* **140**:2513-2529.

Krings, M., Stone, A., Schmitz, R. W., Krainitzki, H., Stoneking, M., and Pääbo, S., 1997, Neanderthal DNA sequences and the origin of modern humans, *Cell,* **90**:19-30.

Kumai, M., and Langway Jr, C. C., 1988, Scanning electron-microscope analysis of aerosols in snow and ice cores from Greenland, *Annals of Glaciology,* **10**:208.

Legrand, M., Fenist-Saigne, C., Saltzman, E. S., Germain, C., and Barkov, N. I., 1991, Ice-core record of oceanic emissions of dimethyl sulphide during the last climate cycle, *Nature,* **350**:144-146.

Liesack, W., and Stackebrandt, E., 1992, Occurrence of novel groups of the Domain *Bacteria* as revealed by analysis of genetic material isolated from an Australian terrestrial environment, *Journal of Bacteriology,* **174**:5072-5078.

Lindahl, T., 1993, Instability and decay of the primary structure of DNA, *Nature,* **362**:709-715.

Lindahl, T., and Nyberg, B., 1972, Rate of depurination of native deoxyribonucleic acid, *Biochemistry,* **11**:3610-3618.

MacAndrews, J. H., 1984, Pollen analysis of the 1973 ice core from Devon Island ice cap, Canada, *Quaternary Research,* **22**:68-73.

Mullis, K. B., and Faloona, F., 1987, Specific synthesis of DNA in vitro via a polymerase-catalyzed chain reaction, *Methods in Enzymology,* **155**:335-350.

Pääbo, S., 1990, Amplifying ancient DNA in: *PCR Protocols: A guide to methods and applications,* M. A. Innis, D. H. Gelfand, J. J. Sninsky and T. J. White, eds., Academic Press, San Diego, pp. 159-166.

Pääbo, S., Higuchi, R. G., and Wilson, A. C. Ancient DNA and the polymerase chain reaction, *Journal of Biological Chemistry.* **264**, 9709-9712 (1989).

Poiner, H. N., Höss, M., Bada, J. L., and Pääbo, S., 1996, Amino acid racemization and the preservation of ancient DNA, *Science,* **272**:864-866.

Rollo, F., Asci, W., Antonini, S., Marota, I., and Ubaldi, M., 1994, Molecular ecology of Neolithic meadow: the DNA of the grass remains from the archaeological site of the Tyrolean Iceman, *Experientia,* **50**:576-584.

Saiki, R. K., Scharf, S., Faloona, F., Mullis, K. B., Horn, G. T., Erlich, H. A., and Arnheim, N., 1985, Enzymatic amplification of β-globin genomic sequences and restriction site analysis for diagnosis of sickle cell anemia, *Science,* **230**:1350-1354.

Saltzman, E. S., Whung, P. Y., and Mayewski, P. A., 1997, Methanesulphonate in the Greenland ice sheet project 2 ice core, *Journal of Geophysical Research,* **102**:26649-26657.

Shaw, K. J., Rather, P. N., Hare, R. S., and Miller, G. H., 1993, Molecular genetics of aminoglycoside resistance genes and familial relationships of the aminoglycoside-modifying enzymes, *Microbiological Reviews,* **57**:138-163.

Schmidt, T., Hummel, S., and Herrmann, B., 1995, Evidence of contamination in PCR laboratory disposables, *Naturwissenschaften,* **82**:423-431.

Speer B. S., Shoemaker, N. B., and Salyers, A. A., 1992, Bacterial resistance to tetracycline: Mechanisms, transfer and clinical significance, *Clinical Microbiology Reviews,* **5**:387-399.

Stone, R., 1999, Permafrost comes alive for Siberian researchers, *Science,* **286**:36-37.

Ubaldi, M., Sassaroli, S., and Rollo, F., 1996, Ribosomal DNA analysis of culturable Deuteromycetes from the Iceman's hay: Comparison of living and mummified fungi, *Ancient Biomolecules,* **1**:35-42.

van Elsas, J. D., 1992, Antibiotic resistance gene transfer in the environment: an overview, in: *Genetic interactions among microorganisms in the natural environment,* E. M. H. Wellington and J. D. van Elsas, eds., Pergamon Press, Oxford, pp. 17-39.

Ward, D. M., Weller, R., and Bateson, M. M., 1990, 16S rRNA sequences reveal numerous uncultured microorganisms in a natural community, *Nature,* **345**:63-65.

Willerslev, E., Hansen, A. J., Christensen, B., Steffensen, J. P., and Arctander, P., 1999, Diversity of Holocene life forms in fossil glacier ice, *Proceedings of the National Academy of Sciences USA,* **96**:8017-8021.

Willerslev, E., Hansen, A. J., Nielsen, K. K., and Adsersen, H. E., in press, Numbers of endemic and native plant species in the Galapagos Archipelago in relation to geographical parameters, *Ecography.*

METEOROLOGICAL AND CLIMATOLOGICAL ASPECTS OF THE SOUTHERN PATAGONIA ICEFIELD

Jorge F. Carrasco[1*], Gino Casassa[2,3], and Andrés Rivera[2,4]

1. ABSTRACT

The Southern Patagonia Icefield (SPI) is located at mid-latitudes in southern South America, which is dominated by the westerly regime and frontal systems. This results in a high frequency of cloudy days (more than 70% of the time) and precipitation events. Analyses of air temperature and precipitation data from southern meteorological stations for the past century indicate an overall warming and decrease in precipitation until the mid-80's, but no significant changes are observed afterwards. In fact, the coastal stations show an increase in precipitation after the 1980's. The mid-term behavior of the atmospheric variables introduces uncertainties in predicting the consequences of future climate change in southern South America.

2. INTRODUCTION

Recent studies indicate that a generalized retreat of glaciers is occurring along the Andes mountains, in particular of those of the Southern Patagonia Icefield (SPI) in southern South America (Figure 1). The SPI is the most glaciated area of the Andes with more than 48 glaciers discharging into fjords on the western side, and into lakes on the eastern side. It is a plateau, located along 73°30' W, between 48 and 51° S, with an area of 13,000 km^2, and an elevation that ranges from 800 to 2000 m with peaks reaching 2500 m, and a few exceeding 3000 m.

SPI is the second largest ice mass in the Southern Hemisphere after Antarctica, and the closest single body of permanent solid water to the equator, being exposed to the influence of the southern Pacific Ocean. Because of the atmospheric environment, the

[1] Dirección Meteorológica de Chile, Casilla 717, Santiago, Chile; [2] Centro de Estudios Científicos (CECS), Arturo Prat 514, Casilla 1469, Valdivia, Chile; [3] Universidad de Magallanes, Casilla 113-D, Punta Arenas Chile; [4] Departamento de Geografía, Universidad de Chile, Marcoleta 250, Santiago, Chile;

* corresponding author: jcarrasco@meteochile.cl

The Patagonian Icefields: A Unique Natural Laboratory for Environmental and Climate Change Studies.
Edited by Gino Casassa et al., Kluwer Academic /Plenum Publishers, 2002.

glaciers that drain from the icefield are of a temperate type with large ablation rates and high flow velocities (Warren and Sugden, 1993).

Several studies indicate that the impact of global warming due to the increase of greenhouse gases observed after the Industrial Revolution can be first detected on the cryosphere where the extension of snow/ice surfaces and glacier outlets respond rapidly to climate changes. The generalized glacier retreat observed in the SPI can be a response to the tropospheric warming detected in the southern tip of South America in the last several decades (Ibarzabal y Donángelo *et al.*, 1996, Rosenblüth *et al.*, 1995). This article presents a summary of meteorological and climatic aspects of southern South America, as well as analyses of recent climatic data, future projection of climate change in the region and the possible impact on the SPI. Available meteorological data from neighboring weather stations and data from the NCEP/NCAR Reanalysis Project are used to characterize the SPI region.

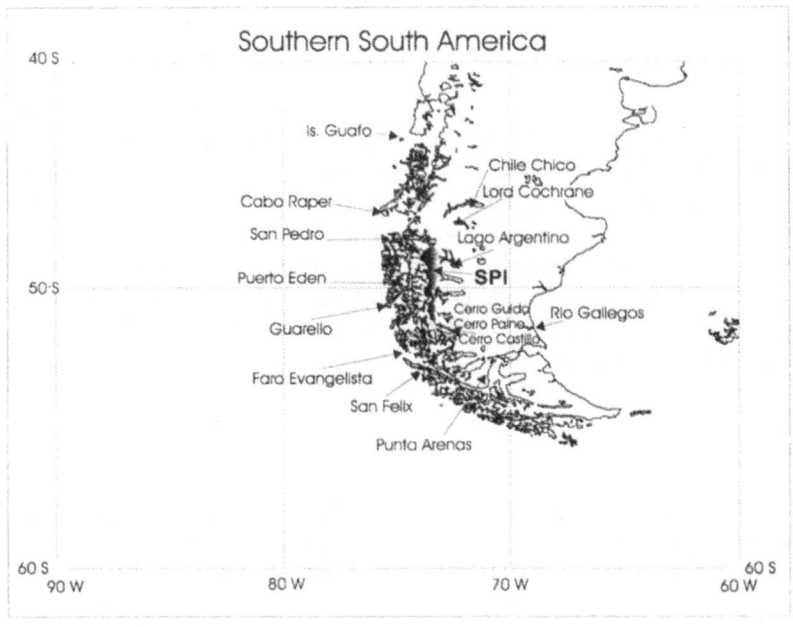

Figure 1. Location map indicating the surface weather stations.

3. METEOROLOGY AND CLIMATOLOGY OF THE SPI

The region is within the mid-latitude westerlies regime as shown by mean wind vectors at 850 hectopascal (hPa) over southern South America for January and July obtained from the NCAR/NCEP Reanalysis (Figure 2). The overall annual behavior indicates that the winds are weaker during the winter, with a southwesterly direction compared to summer, when the wind regime becomes stronger and with a north-northwesterly direction. Analyses of frontal and cyclone trajectories, reveal that the SPI is frequently affected by these weather systems. This results in a high number of cloudy days and precipitation events throughout the year. For example, the coastal station San Pedro, located near the northern end of the SPI (47°43' S, 74°55' W), has an annual average of 77% daily cloud cover, and 297 days with precipitation. Faro Evangelistas, located about 150 km south of the SPI (52°24' S, 75°36' W), registers an even larger annual average of 86% daily cloud cover and 322 days with precipitation.

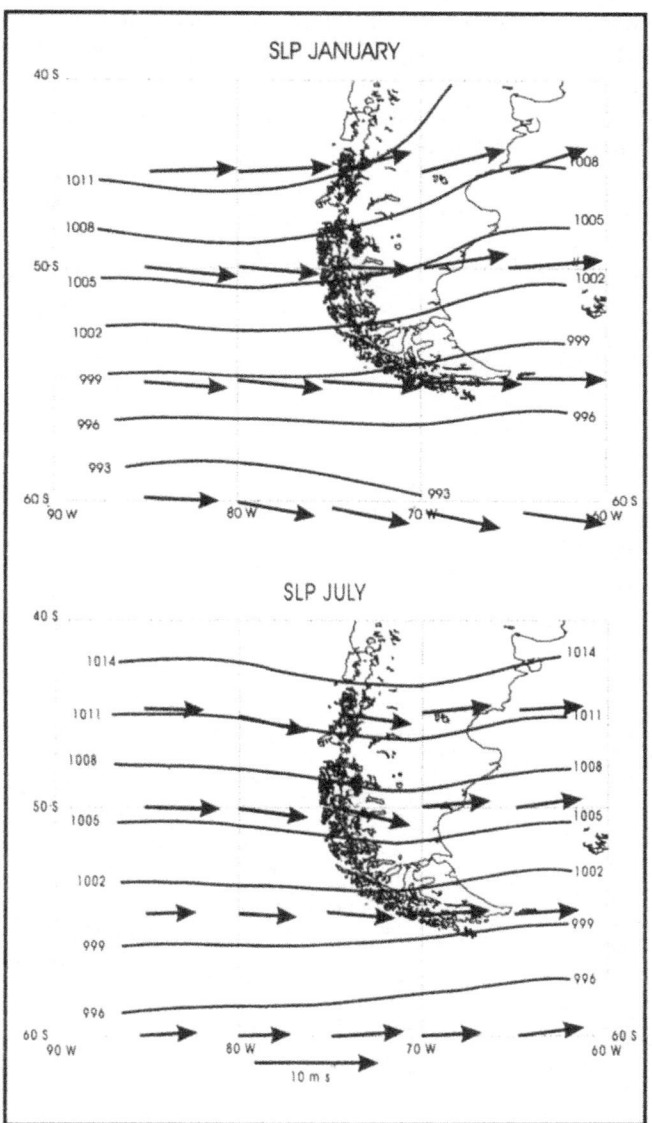

Figure 2. Mean sea-level pressure isobars (in hectopascals: hPa) along with average wind vectors at 850-hPa (around 1350 m asl) for January and July.

The precipitation regime, however, is affected by the west-east topographic profile of southern South America. This introduces a spatial distribution with a higher amount of precipitation on the west side, with an annual amount that exceeds 7000 mm on the coast and 10,000 mm on the icefield (DGA, 1987). These amounts rapidly decrease toward the east in the Argentine Patagonia region, where the annual precipitation, on average, is below 400 mm (Carrasco *et al.*, 1998, Ibarzabal y Donángelo *et al.*, 1996). Figures 3a and 3b show the average monthly-accumulated precipitation at several weather stations located on the west and east sides of the SPI. The almost straight lines indicate the homogeneous distribution of the precipitation throughout the year. It can be noted that while all the western stations show a total annual precipitation above 2000 m, all the eastern stations show a value of less than 700 mm, and actually 4 out of 6 have a total annual accumulation of under 400 mm.

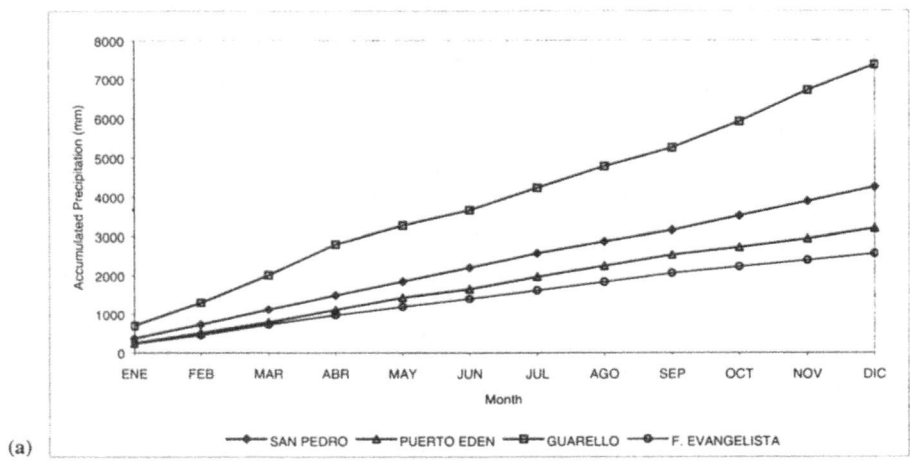

(a)

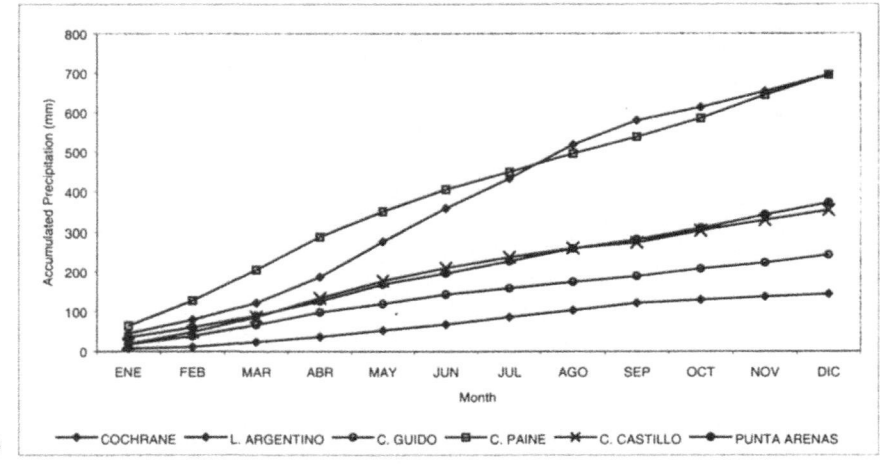

(b)

Figure 3. Mean monthly precipitation accumulation for stations located on the western (a) and eastern (b) side of the Southern Patagonia Icefield.

The temperature behavior also reveals a slight difference between coastal stations, influenced by the Pacific Ocean, and the inland stations influenced by the continental environment. Figure 4 depicts the monthly average temperature at six stations. Overall, stations located on the east side of the SPI (Cochrane, Lago Argentino, and Punta Arenas) show larger seasonal oscillations with warmer (colder) temperature during the summer (winter) than the western stations (San Pedro, Puerto Edén and Faro Evangelistas).

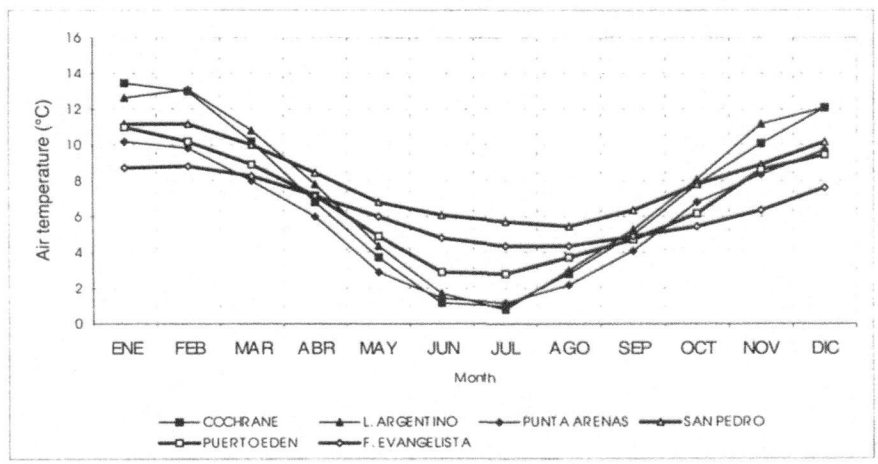

Figure 4. Mean monthly surface air temperature for stations located on the western (thin line) and eastern (thick line) side of the Southern Patagonia Icefield.

4. RECENT CLIMATE VARIABILITY

4.1. Temperature

The near-surface linear warming during the last 100 years in the vicinity of the SPI is between 1.3 and 2.0 °C based on a homogenized period between 1933 and 1992 (Rosenblüth *et al.*, 1997). At Faro Evangelistas, the total warming is about 0.7 °C during the period 1901-1988 (Rosenblüth *et al.*, 1995), and at Río Gallegos, a warming of 1.4 °C from 1938 to 1988 was registered (Ibarzabal y Donángelo *et al.*, 1996). In general, south of 45° S a positive trend (warming) is revealed at stations located in southern South America, on both sides of the SPI.

A comparison was made between the 850 hectopascal geopotential height (hPa), and other levels, obtained from radiosondes (1975 onward) at Punta Arenas and from nearby grid points of the NCEP/NCAR reanalysis (at 52.5° S, 70° W). Punta Arenas is the closest radiosonde station to the SPI and the atmospheric variables measured at this location can be used as a reference for analyzing the meteorological behavior and changes on the icefield. A value of 850-hPa is used because it is about 1350 m above sea level in the free atmosphere, which concurs with the average elevation (1336 m) of the whole icefield (Casassa and Rivera, 1999). Results indicate a correlation coefficient of 0.98 between the observed temperature at Punta Arenas and the derived temperature from

NCEP/NCAR reanalysis, suggesting that the re-analyses dataset can provide 50 years (to be extended in the future) for examining atmospheric changes during this period. Figures 5 and 6, respectively, show the monthly and annual mean temperature at 850-hPa obtained by the radiosonde measurements and those derived from the re-analysis.

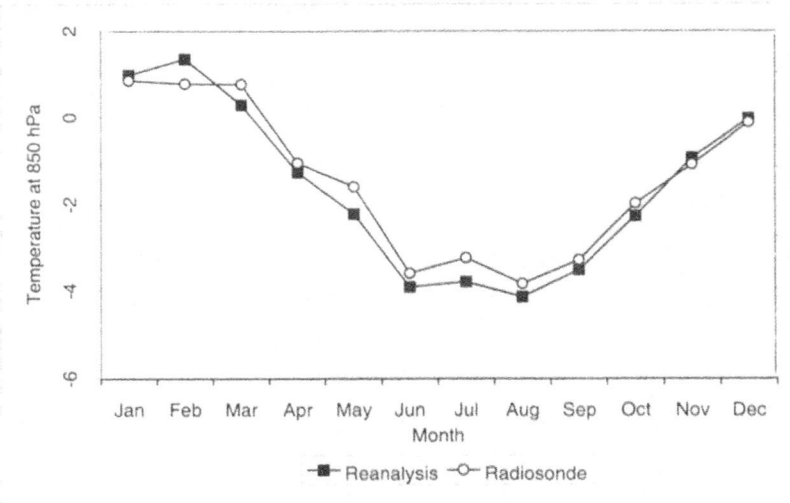

Figure 5. Monthly average behavior of the air temperature at 850-hPa, recorded by the radiosonde at Punta Arenas and derived from the reanalysis for the 1977-1998 period.

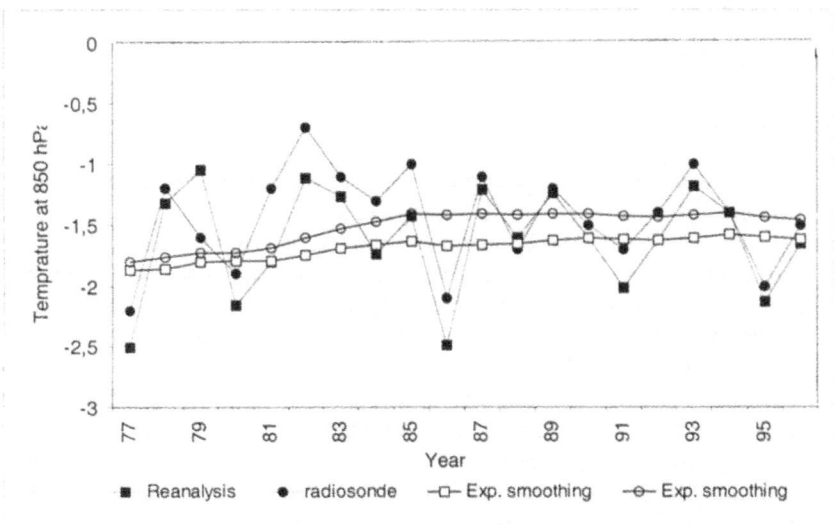

Figure 6. Annual means of air temperature at 850-hPa for the 1977-1998 period, recorded by the radiosonde at Punta Arenas and derived from the reanalysis.

In Figure 6, the thin lines with large dots and squares indicate annual mean while the thick lines, with smaller dots and squares represent the twice-exponential smoothing (see Rosenbluth *et al.*, 1995, 1997) given by,

$$y_t = cx_t + (1-c)y_{t-1} \qquad t = 2, \ 3,\ldots n$$

for the first forward smoothing, where the first value y_l is given by the average of the first 7 years of the series, and

$$z_t = cy_t + (1-c)z_{t+1} \qquad t = (n-1), \ (n-2),\ldots 1$$

for the second backward smoothing in time, with z_t (the last value) assumed to be the final y_t value of the first smoothing.

The coefficient c is the degree of smoothing applied to the data that range from 0 (maximum smoothing) to 1 (no smoothing). Here, a value of $c = 0.25$ is used so that the filtering does not smooth out important interannual variability (see Rosenblüth *et al.*, 1995). It can be seen in Figures 5 and 6 that both datasets describe a very similar seasonal and interannual behavior of the air temperature at 850-hPa level. However, the real radiosonde data reveal higher warming than the reanalysis. Actually, the warming is significant until 1985 and afterwards no change is observed at all. Figure 7 extends back the series until 1958 using the reanalysis data. It reveals an overall warming of about 0.98 °C since 1958, with a significant increase until the mid-1970's and then a slight increase until 1985. This trend is also seen in Figure 6.

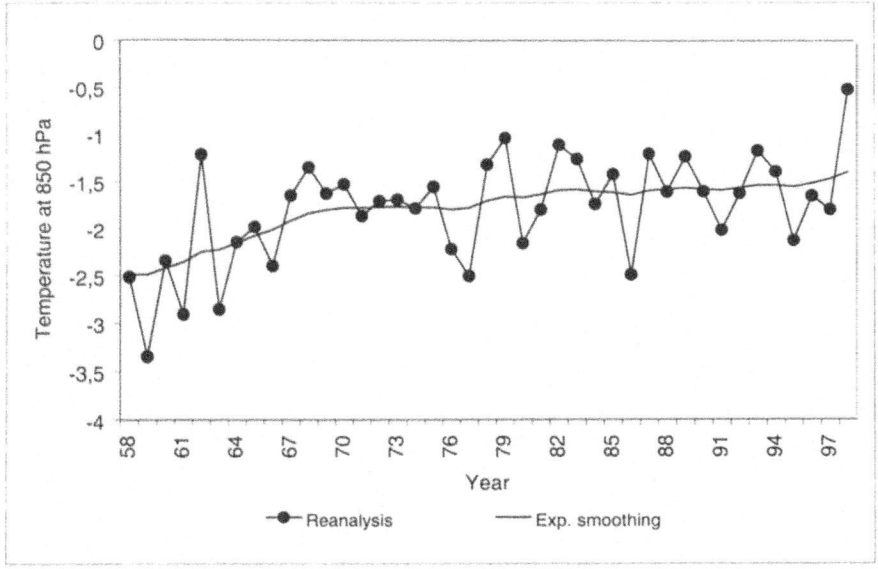

Figure 7. Extended annual mean of the air temperature at 850-hPa derived from the reanalysis dataset at the closest grid point from the SPI.

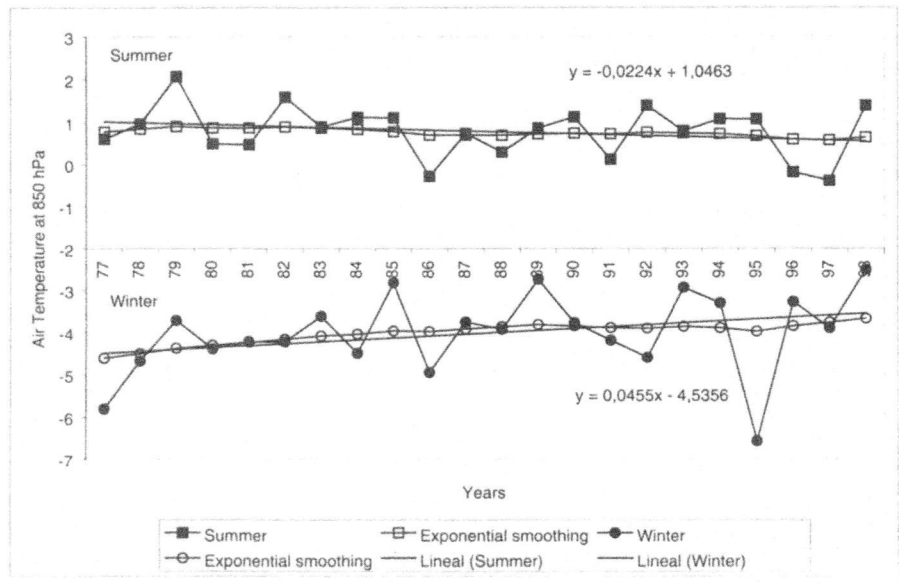

Figure 8. Mean summer (December-January-February) and winter (June-July-August) air temperatures at 850–hPa obtained from the reanalysis at the closest grid point from the SPI.

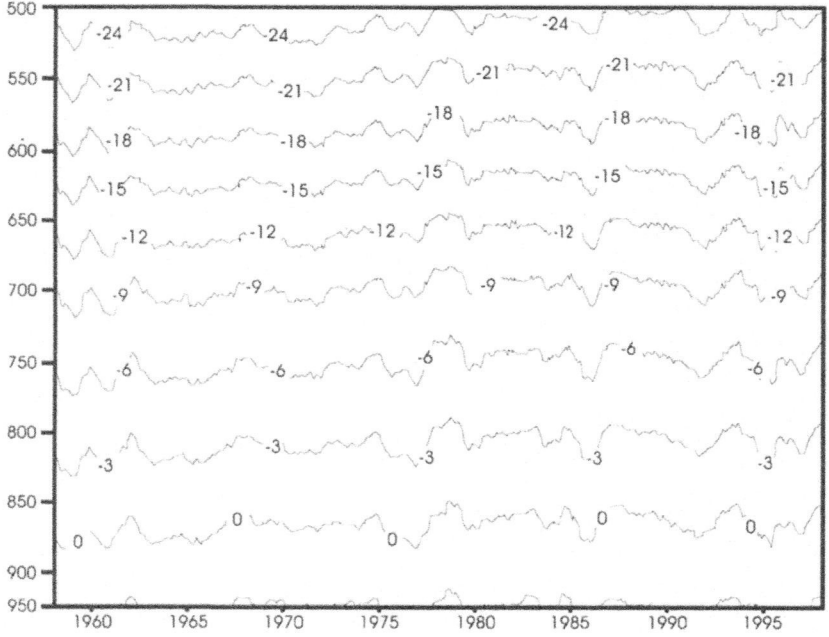

Figure 9. Air temperature behavior at the lower troposphere derived from the re-analyses at the closest grid point from the SPI. Values within the figure are expressed as °C.

Figure 8 shows the 850-hPa reanalysis of air temperature at Punta Arenas for summer and winter from 1977 onward. The linear trend for summer reveals a cooling (-0.0224 °C yr^{-1}) of about -0.45 °C in 22 years, while in winter there is a warming of about 1 °C in the same period (0.0455 °C yr^{-1}). This indicates that the observed overall warming at 850-hPa is due to the winter warming.

The high correlation between reanalysis and observed radiosonde data allows us to use the reanalysis for exploring the atmospheric behavior of the SPI. Thus, Figure 9 depicts the temperature behavior in the lower half of the troposphere, obtained from the reanalysis data at the closest grid point to the SPI (at 50° S, 75° W). Once again, a large interannual and seasonal variability can be observed, as well as a slight overall warming at all levels. The zero isotherm level fluctuates between 1100 m in winter and 1400 m in summer, which means that during the winter, more than 90% of the SPI is above the zero isotherm level and during the summer, about half of the SPI is above the zero isotherm level. The equilibrium line elevation for the SPI ranges between 900 and 1250 m (Casassa and Rivera, 1999), which is close to the winter zero isotherm elevation.

4.2. Precipitation

Precipitation records over the southern tip of South America are not complete because gaps exist during some months and even a few complete years are missing. Rosenblüth et al., (1995) analyzed the available and incomplete raw data set with an exponential smoothing and found an overall decrease ranging from 25% to 33% in Faro Evangelistas and Bahía Felix (53.0° S, 74.1° W) during 1899-1988 and 1915-1985, respectively. Punta Arenas showed an interannual oscillation of about 40 years with maximum precipitation about 1915 and 1950, but did not reveal a significant change between the mid-1960s and 1988. Over the eastern side, Lago Argentino (50°20' S, 72°18' W) and Río Gallegos (51°37' S, 69°17' W) stations showed a negative trend during 1937-1990 and 1927-1990, respectively (Ibarzabal y Donángelo et al., 1996).

Figures 10a and 10b show the precipitation data of Rosenblüth et al. (1995) for Punta Arenas, Faro Evangelistas and Isla Guafo up to 1998. Unfortunately, Bahía Felix only operated until 1988. Figure 10a also includes San Pedro, Faro Cabo Raper (46°50' S, 75°35' W), Chile Chico and Lord Cochrane, as reference stations for analyzing precipitation on the northern part of the SPI. For these last two stations, only about 30 years of data are available. The monthly (annual) average of the two previous and subsequent months (years) and the respective long-term monthly (annual) average were used to complete missing months (years) where possible. This arbitrary procedure is performed in order to apply the exponential smoothing and therefore, results must be taken with caution.

A striking increase of precipitation since about 1983 is observed at Faro Evangelistas. The weather station at this site has not changed location nor instrument type according to the Servicio Meteorológico de la Armada (personal communication), which means that the significant precipitation increase could be real, although its magnitude should be taken with caution. Faro Evangelistas is located just south of a region of maximum precipitation (Miller, 1976; Prohaska, 1976) so it can capture any slight change in storm trajectory. Further analysis is needed in the future.

Besides the findings described by Rosenblüth et al. (1995) up to 1988, the updated southernmost stations Faro Evangelistas and Punta Arenas revealed a precipitation increase since the mid-1980s onward. The same can be seen for Isla Guafo and in the

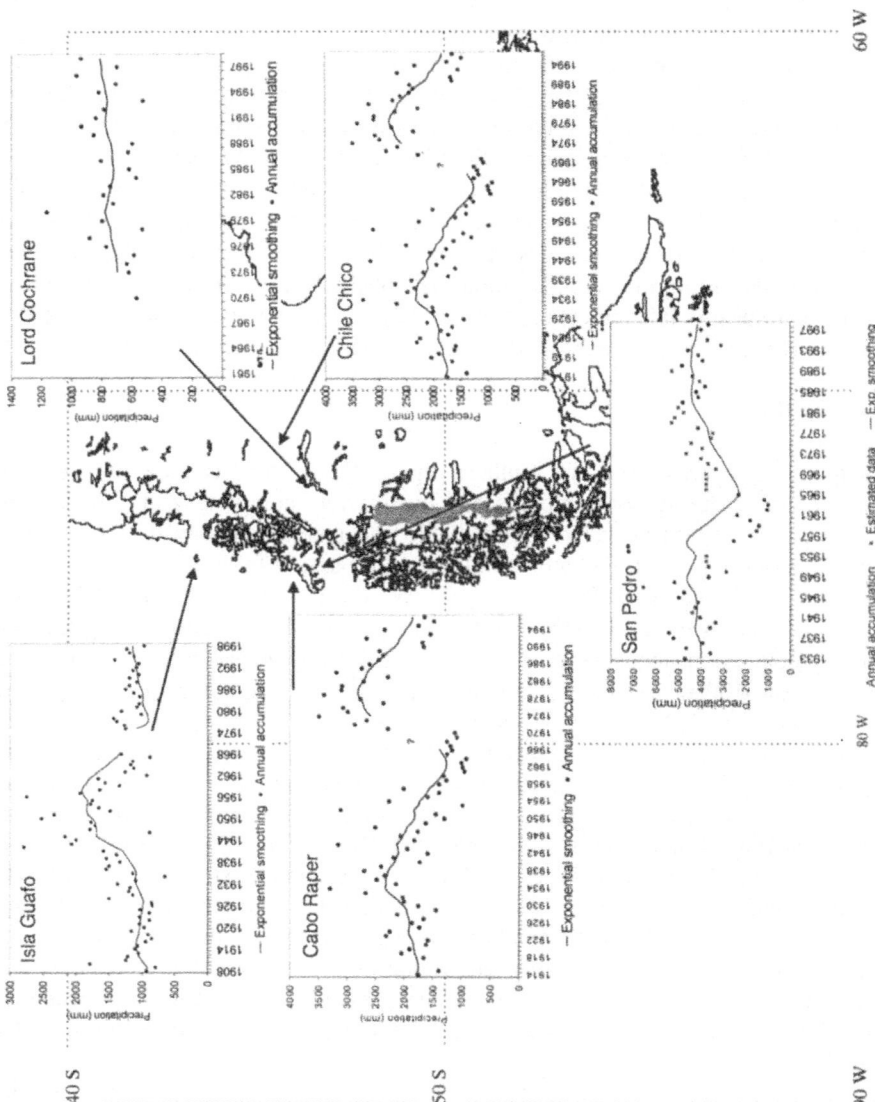

Figure 10a. Precipitation records in the Chilean stations located in southern South America - northern end.

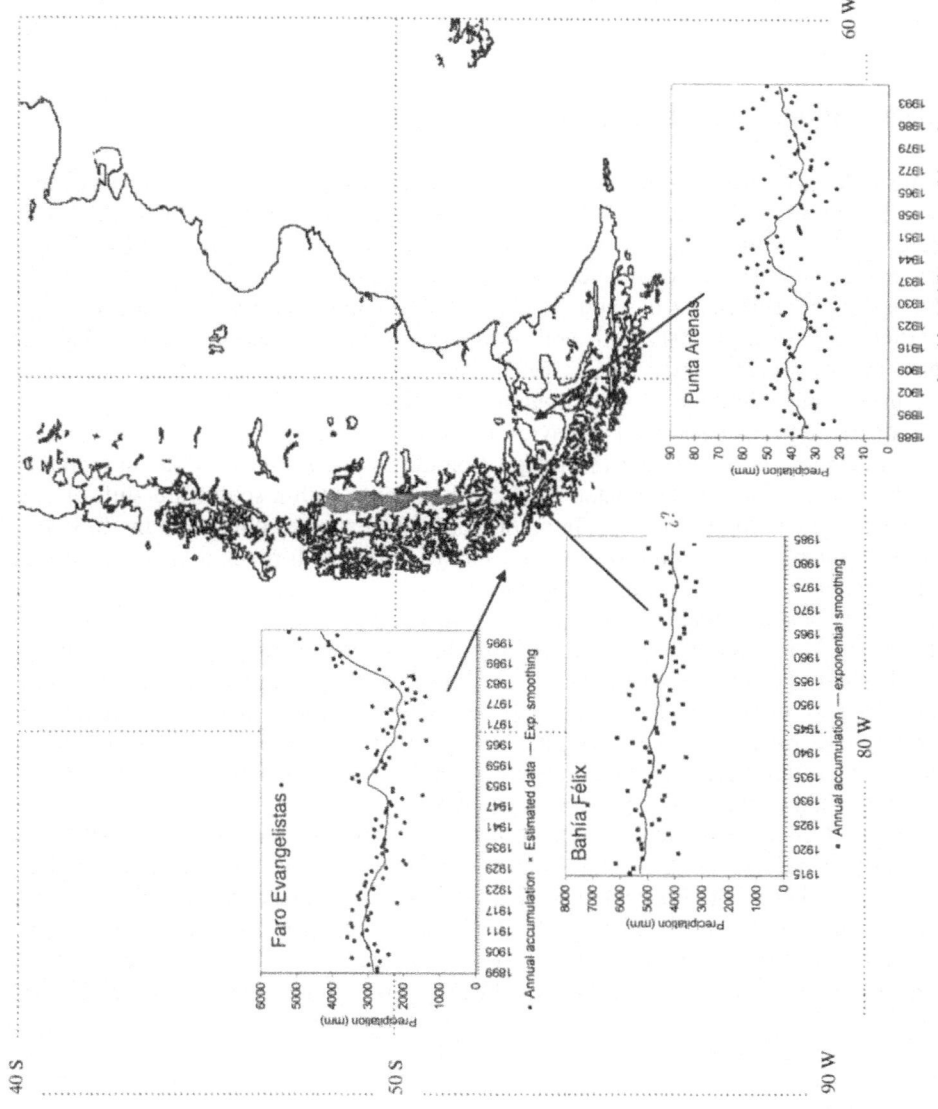

Figure 10b. Precipitation records in the Chilean stations located in southern South America - southern end.

short records of Chile Chico and Lord Cochrane stations. However, Cabo Raper and San Pedro stations showed a general decrease after the mid-1980s. This indicates a high spatial variability of precipitation over the region and no clear conclusion can be interpreted from the available data with respect to significant changes during the last century.

5. DISCUSSION AND CONCLUSION

Many global climate model simulations with double CO_2 have been applied since the 1980's. The results have suggested that the air temperature in the region of the SPI will increase between 2 and 6 °C in summer and between 2 and 4 °C in winter (IPCC, 1990; 1992), by 2030-2050. Later, another model that included the aerosol effects (IPCC, 1995), suggested a warming of about 2 and 3 °C for summer and winter, respectively, from 1880-1889 to 2040-2049. All model simulations with increased CO_2 revealed an increase in global precipitation (IPCC, 1995). However, because precipitation still has a high degree of uncertainty in climate simulation, caution is required when interpreting the spatial distribution and seasonal behavior of this parameter. Overall, model results suggest a decrease in the amount of precipitation during the summer, while an increase should be expected during the winter. This can be related with a southward shift of the mean frontal depression trajectories. In other words, the expected climate in the SPI will evolve to resemble the conditions observed farther north, with precipitation showing a seasonal behavior, and with warmer temperatures.

A review of the available historical data indicates an overall warming and decrease in precipitation in southern South America during the last century. However, climate trends changed after the mid-1980s and in fact, an important increase in precipitation is apparent at several stations during the last 15-20 years (Figure 10). This reveals the uncertainty in predicting the precipitation behavior and the need for a better understanding of the expected changes in the general climate circulation and their impact on the spatial and temporal precipitation distribution.

In light of the climatic trends observed in the last few decades until the mid-1980's, that is, the overall warming, together with the average negative trend in precipitation observed around the SPI (Rosenblüth *et al.*, 1995, 1997; Ibarzabal y Donángelo *et al.*, 1996), are thought to be the most probable cause for the generalized glacier retreat observed in southern South America. However, the relative cooling and precipitation increase observed after the mid-1980's could result in regional glacier advance in the near future.

6. ACKNOWLEDGMENTS

This work is a contribution of the Dirección Meteorológica de Chile (DMC) and it is partially supported by the project Fondecyt No 1980293. Data belong to, and were made available by, the DMC and the Servicio Meteorológico de la Armada de Chile. Comments by Al Rasmussen are greatly appreciated.

7. REFERENCES

Carrasco, J. F., Casassa, G., and Rivera, A., 1998, Climatología actual del Campo de Hielo Sur y posibles cambios por incremento del efecto invernadero, *Anales Instituto de la Patagonia*, Serie Ciencias Naturales, **26**:119-128 (in Spanish).

Casassa, G., and Rivera, A., 1999, Topographic mass balance model for the Southern Patagonia Icefield, Abstracts International Symposium on the Verification of Cryospheric models, Bringing Data and Modeling Scientists Together, 16-20 August 1999, Zurich, p. 44.

DGA, 1987, Balance Hídrico de Chile, Dirección General de Aguas, Santiago, Chile (in Spanish).

Ibarzabal y Donángelo, T., Hoffmann, J. A. J., and Naruse, R., 1996, Recent climate changes in southern Patagonia, *Bulletin of Glacier Research*, **14**:29-36.

IPCC, 1990, *Climate Change, The IPCC Scientific Assessment*, J. T. Houghton, G. J. Jenkins and J. J. Ephraums, eds., Cambridge University Press, Cambridge, UK, pp. 365.

IPCC, 1992, *Climate Change 1992: The Supplementary Report to the IPCC Scientific Assessment*, J. T. Houghton, G. J. Jenkins and J. J. Ephraums, eds., Cambridge University Press, Cambridge, UK, pp. 365.

IPCC, 1995, *Climate Change 1995: The Science of Climate Change*, J. T. Houghton, L. G. Meira Filho, B. A. Callander, N. Harris, A. Kattenberg and K. Maskell, eds., Cambridge University Press, Cambridge, UK, pp. 572.

Miller, A., 1976, The Climate of Chile, in: *Climate of Central and South America*. World Survey Climatology Volume 12, W. Schwerdtfeger, ed., Elsevier Scientific Publishing Company, pp. 113-145.

Prohaska, F., 1976, The Climate of Argentina, Paraguay and Uruguay, in: *Climate of Central and South America*. World Survey Climatology Volume 12, W. Schwerdtfeger, ed., Elsevier Scientific Publishing Company, pp. 13-112.

Rosenblüth, B., Casassa, G., and Fuenzalida, H., 1995, Recent climatic changes in western Patagonia, *Bulletin of Glacier Research*, **13**:127-132.

Rosenblüth, B., Fuenzalida, H. A., and Aceituno, P., 1997, Recent temperature variations in Southern South America, *International Journal of Climatology*, **17**:67-85.

Warren, C., and D. Sugden, 1993, The Patagonian icefields: a glaciological review, *Arctic and Alpine Research*, **25**(4):316-331.

WESTERN PATAGONIA: A KEY AREA FOR UNDERSTANDING QUATERNARY PALEOCLIMATE AT SOUTHERN MID-LATITUDES

Patricio I. Moreno[1*]

1. ABSTRACT

Paleoclimate research in the Andean region south of 40° S is key to understanding the climatic evolution of the mid-latitudes of the Southern Hemisphere. Studies in this region are of pivotal importance for determining the mechanisms responsible of climate changes at regional, hemispheric, and global scales during the Quaternary. This monumental task requires a multidisciplinary approach to understanding the modern environments (meteorology, glaciology, oceanography, modern biota) on and around the Patagonian icefields, and past variations in climate through the development of multiple climate proxies (stratigraphic analysis of ice and sediment cores, paleoecological studies, tree ring chronologies, among others).

2. THE PROBLEM

2.1. Introduction

The causes and consequences of Ice Ages throughout the Quaternary, i.e. the last 1.8 million years, pose a major scientific question. Periodic ice ages have dominated the Quaternary, punctuated by relatively brief periods of interglacial conditions. The present interglacial, the Holocene, started about 11,000 years ago and represents the stage at which agriculture and human civilizations began and developed.

A striking feature of Quaternary ice ages is the occurrence of cyclic, large-scale changes in the ocean, atmosphere, and the cryosphere. Studies in the 1920s and 1930s, by the Yugoslavian astronomer Milutin Milankovitch, led to the idea that past changes in summer radiation at high northern latitudes drove the climate system into periodic ice

[1] Patricio I. Moreno, Departamento de Biología, Universidad de Chile, Las Palmeras 3425, Santiago, Chile

* corresponding author: pimoreno@uchile.cl

The Patagonian Icefields: A Unique Natural Laboratory for Environmental and Climate Change Studies.
Edited by Gino Casassa et al., Kluwer Academic /Plenum Publishers, 2002.

ages (Broecker and Denton, 1989). Milankovitch calculated past variations in summer insolation at 65° N, considering the additive effects of three independent movements of the Earth: eccentricity (changes in the elliptical shape of the Earth's trajectory around the sun), obliquity (changes in the tilt of the Earth's axis), and precession of the equinoxes. The correspondence between key paleoclimate events and the insolation curves produced by Milankovitch and subsequent workers, lent support to the idea that past variations in the radiative budget of the Earth - driven by astronomic mechanisms- were the pacemaker of the Ice Ages (Broecker and Denton, 1989; Hays *et al.*, 1976; Shackleton, 2000). Spectral analyses on the calculated insolation values have revealed the dominance of the 19,000/23,000-year precession and 41,000-year obliquity cyclicities, with a relatively minor contribution from the 100,000-year periodicity (eccentricity). In contrast, analyses of marine isotope records spanning several glacial cycles have revealed a dominant 100,000-year periodicity for the succession of ice ages, accompanied by less-dominant cycles of 19,000/23,000 and 41,000 years (Broecker and Denton, 1989). The 100,000-year periodicity is one of the most enigmatic aspects of Quaternary Ice Ages, because it is not a straightforward consequence of the astronomic mechanism. Recent papers on this subject have stressed the role of complex feedback mechanisms linked to the greenhouse effect of atmospheric carbon dioxide (Shackleton, 2000), and an asymmetric thermohaline-ice sheet oscillator (Denton, 2000).

Although the astronomical theory is a widely accepted paradigm in the paleoclimate community, research in past decades has raised questions as to (a) whether insolation is the sole or primary mechanism driving paleoclimate change, and (b) how insolation variations trigger reorganizations in the climate system at hemispheric and global scales. Three prominent observations have challenged the notion of Milankovitch orbital forcing as the key driving mechanism: (i) the predominance of a 100,000-year cycle in paleoclimate records, (ii) the identification of abrupt climate changes at sub-Milankovitch time scales, such as Dansgaard-Oescher, Heinrich events and related cycles (Behl and Kennet, 1996; Bond *et al.*, 1992; Bond and Lotti, 1995; Broecker, 1994; Broecker *et al.*, 1992; Grootes *et al.*, 1993; Heinrich, 1988) and (iii) the abruptness of ice age terminations and oceanic turnover shifts (Broecker 1991; Broecker 1995; Broecker, 1997; Broecker and Denton, 1989; Lehman and Keigwin, 1992) . Paleoclimate records showing these high-frequency fluctuations have been developed, for the most part, in the Northern Hemisphere, and it remains to be demonstrated whether these signals represent climate change on a regional or a global scale. To gain a broader understanding of the causes and consequences of these millennial-scale fluctuations it is necessary to determine their geographic extent, particularly in regions where climate controls and orbital forcing are most contrasting. Good examples of abrupt changes in the ocean-atmosphere system, unrelated to Milankovitch forcing, have been reported during the termination of the last ice age, i.e. 14,600-10,000 ^{14}C yr B.P. (Alley *et al.*, 1993; Atkinson *et al.*, 1987; Bard *et al.*, 1987; Björk *et al.*, 1998; Dansgaard *et al.*, 1989; Mayewski *et al.*, 1993; Mayewski *et al.*, 1996). Concurrent with the onset of the last termination, sedimentary records from the North Atlantic Ocean reveal massive discharge of icebergs exiting the Laurentide and Scandinavian ice sheets, during the so-called Heinrich events (Andrews 1998; Bond *et al.*, 1992; Bond and Lotti, 1995; Heinrich, 1988). These events caused abrupt changes in sea surface temperature, salinity, and deep-water formation, which in turn, locked the amphi-North Atlantic region into glacial conditions. The Bölling-age (13,000 ^{14}C yr B.P.) warming in the amphi-North Atlantic region was an extraordinarily rapid event, temperature rose from glacial maximum to near-modern values in less than two hundred years. Another key event in the deglaciation process in the Northern Hemisphere was the Younger Dryas event (11,000-10,000 ^{14}C yr B.P.; Björk et al., 1998; Walker, 1995). Paleoclimate records from Europe and Greenland ice cores show an abrupt reversal to glacial maximum conditions during Younger Dryas time, following the establishment of conditions similar to modern climate during the Bolling-Allerod interstadial. Climate transitions during the Younger Dryas took place in a matter of a few decades, as indicated by annually-resolved ice-core records (Alley et al., 1993).

Unraveling the paleoclimate history of western Patagonia (40°-50° S) is essential for determining whether climatic changes in the Northern and Southern Hemispheres were synchronous, or if variations in one hemisphere followed those in the other hemisphere and hence, to better understand the underlying mechanisms operative during ice-age cycles. Likewise, comparisons between Antarctic paleoclimate records and data from western Patagonia will help us understand ice-age climate dynamics at a hemispheric level, and the relative contributions made by low- and high-latitude climate processes, to regional climate changes.

2.2. The Southern Andes as a Key Region for Paleoclimate Research

Very few land areas in the mid-latitudes of the Southern Hemisphere are suited for the study of the ice-age history of the southern westerly winds and of the timing, rates and direction of climate changes during the Quaternary. One of these areas is the Andean region, south of 40° S, because it is particularly sensitive to the influence of the southern westerlies, and it is one of the few land areas sustaining icefields, alpine glaciers, and rain forest communities along altitudinal and latitudinal gradients within the belt of the southern westerlies. An important advantage of conducting studies in this area is the abundance of organic material incorporated in glacial, lacustrine, and wetland deposits, which makes it possible to develop stratigraphic studies controlled by precise radiocarbon chronologies. Therefore, detailed paleoclimate records along a latitudinal transect can be used to track the ice-age history of the southern westerlies, and the anatomy of climate changes during and since the last ice age.

Western Patagonia is a key area for testing hypotheses of global climate change, stressing the role of:

1. Orbital-induced changes in insolation, because seasonal insolation regimes in the polar hemispheres are out-of-phase.
2. Changes in deep-water production in the North Atlantic and Southern Oceans. Based on the δ^{14}C record from the Cariaco basin during Younger Dryas time (Hughen et al., 1998) and the differences in timing and direction of climate changes during the last deglaciation in Greenland and Antarctica (Sowers and Bender, 1995), Broecker (1998) proposed that a bipolar see-saw in deep water production between the North Atlantic Ocean and the Southern Ocean, caused out-of-phase (antiphase) climate changes in the polar hemispheres.
3. Internal ice-sheet dynamics. The southern mid-latitudes are predominantly oceanic, are far removed from the Northern Hemisphere ice sheets, and lacked large ice sheets during the last glaciation, which may have masked or triggered regional climate changes.

4. The role of the southern westerlies as a key component of the global climate system. Imbrie *et al.*, (1992), and Toggweiller and Samuels (1995), proposed that variations in the Southern Hemisphere westerlies may have an impact on the amount of carbon dissolved in the Southern Ocean, the extent of Antarctic sea ice, and global thermohaline circulation. This influence is due to the fact that the wind stress imparted by the westerlies on the Southern Ocean at, and south of, the latitude of Drake Passage (56°-60° S Lat.), forces a northward flow of surface waters which, in turn, drives deep thermohaline flow into the Antarctic region. The varying efficiency of this mechanism through an ice-age cycle is linked to the intensity and latitudinal position of the southern westerlies.

3. PALEOCLIMATE RECORDS FROM WESTERN PATAGONIA

3.1. Last Glacial Maximum (LGM) and Deglaciation (Last Termination)

Late Quaternary sedimentary deposits of the Región de los Lagos and Isla Grande de Chiloé (39-43° 30' S), have been the focus of intensive palynological and glacial-geological research over the past 35 years (Heusser, 1966; Heusser, 1974; Heusser, 1981; Heusser, 1984; Heusser, 1989; Heusser, 1990; Heusser and Flint, 1977; Heusser *et al.*, 1999; Heusser *et al.*, 1996a; Heusser *et al.*, 1996b; Laugenie, 1982; Mercer, 1976; Porter, 1981). On at least four occasions, piedmont glacier lobes descended from the Andes to the Valle Longitudinal of Región de los Lagos (Mercer, 1976; Porter, 1981), with progressively smaller maxima, apparently as a consequence of tectonic subsidence (Clapperton, 1991a). The timing and extent of the older glaciations is poorly understood. Deposits of the last ice age, called the Llanquihue Glaciation by Heusser (1974), have been studied in detail in this region; these deposits represent the type sequence for the last glaciation in South America (Clapperton, 1991b). Based on the well-preserved geomorphology and the available radiocarbon chronology of the glacial deposits, Porter (1981) recognized three maxima for the Llanquihue and Seno Reloncaví ice lobes during the Llanquihue Glaciation. More recently, Lowell *et al.* (1995) and Denton *et al.* (1999a; 1999b), extended the radiocarbon chronology of the Llanquihue-age glacial deposits, recognizing at least four glacial maxima during the Llanquihue Glaciation between 29,400-14,600 ^{14}C yr B.P. These studies indicate that glaciers collapsed during the ~14,600 ^{14}C yr B.P. event in Región de los Lagos and Isla Grande de Chiloé, and receded deep into the Cordillera de los Andes. Several pollen sites from Región de los Lagos and Isla Grande de Chiloé yielded ages between 13,000-12,000 ^{14}C yr B.P. for the replacement of *Nothofagus* parkland and Subantarctic taxa by North Patagonian rainforests, implying climate amelioration at the end of the last glaciation (see below). Fossil-beetle studies from the same area indicate climate warming at about 14,000 ^{14}C yr B.P. (Hoganson and Ashworth, 1992), in agreement with glacial geologic records.

Studies on the glacial history of the channels region (45°-50° S) indicate that glaciers had receded from their LGM position in Península de Taitao and Archipiélago de los Chonos (45°-47° S) by ~14,335 ^{14}C yr B.P. (Bennett *et al.*, 2000; Lumley and Switsur, 1993), and from Puerto Edén (49°S) by ~13,000 ^{14}C yr B.P. (Ashworth *et al.*, 1991). The available chronologies indicate that glaciers in this region did not re-advance during Younger Dryas time (Ashworth *et al.*, 1991; Mercer, 1976). Studies near Lago

Viedma (Wenzens, 1999) and Torres del Paine (Marden, 1997; Marden and Clapperton, 1995), indicate pulses of glacial expansion during late-glacial time; however, the timing of those events is poorly constrained, and their regional significance has not been demonstrated.

Glacial geologic studies in the Estrecho de Magallanes area (53°-55° S) show that glacial recession occurred during the last termination, prior to 14,260 ^{14}C yr B.P., and was followed by a re-advance between 12,700-10,300 ^{14}C yr B.P. (McCulloch et al., 2000). The latter event seems to be related to the southward migration of the westerlies from their glacial maximum position, and the resulting net increase in mass balance of glaciers in this region (see below).

The latitudinal position of the Southern Hemisphere westerlies during the last glaciation has been the focus of controversy, centered on whether the storm tracks shifted northward (Heusser, 1989) or whether they were intensified and focused year-round south of 42° S Lat. (Markgraf, 1989). The origin of this controversy lies, in part, in the climatic scenario for the Región de los Lagos during the portion of the LGM between 20,000-14,000 ^{14}C yr B.P. Studies based on pollen records (Heusser, 1974; Heusser, 1981) suggest a cold, dry environment existed; however, there are disagreements on the mechanisms to account for decreased precipitation in the Región de los Lagos during the LGM. Markgraf (1989) proposed that increased aridity resulted from a poleward shift of the northern margin of the westerlies to latitudes south of 42° S Lat. In contrast, Heusser (1974) alluded to decreased offshore evaporation caused by the intensification of the cold Humboldt Current. Subsequently, Hoganson and Ashworth (1992) argued for cold, wet climate conditions during the LGM, based on fossil beetle assemblages from the Región de los Lagos.

3.2. Late-Glacial Climate and the Younger Dryas Problem

Discrepancies about the timing, frequency, and direction of climate changes between 14,600-10,000 ^{14}C yr B.P. have led to divergent views on the pattern of climate change during late-glacial time. One model, based on palynological and paleoentomological records (Hoganson and Ashworth, 1992), proposes a single-step warming at ~14,000 ^{14}C yr B.P. during the last termination. Another model, based on pollen records from the Región de los Lagos and Isla Grande de Chiloé, recognizes a warming event at about 12,500 ^{14}C yr B.P., followed by climate cooling starting at 11,000 ^{14}C yr B.P. (Heusser, 1966; Heusser, 1981; Heusser et al., 1981). Subsequent studies by Calvin Heusser proposed a warming event at about 13,900 ^{14}C yr B.P. and one (several?) cooling event(s), starting at 13,000 ^{14}C yr B.P. (Heusser et al., 1995; Heusser et al., 1996a; Heusser et al., 1996b) and/or 12,000 ^{14}C yr B.P. and/or 11,000 ^{14}C yr B.P. (Heusser, 1993). Recent studies (Bennett et al., 2000; Lumley and Switsur, 1993; Moreno, 1997; Moreno et al., 1999) have shown that North Patagonian rainforest taxa expanded following ice recession at 14,600-14,300 ^{14}C yr B.P. in Región de los Lagos, Isla Grande de Chiloé, Archipelago de los Chonos, and Península de Taitao. Subsequent climate changes starting at ~10,000 ^{14}C yr B.P. completed the temperature and precipitation recovery to full interglacial conditions.

The presence or absence of a cooling event of Younger Dryas age (11,000-10,000 ^{14}C yr B.P.) is one of the most debated topics in South American paleoclimatology. Putative pollen evidence in support of a Younger Dryas-age cooling from the Región de los Lagos and Isla Grande de Chiloé (Heusser, 1993; Heusser *et al.*, 1995; Heusser *et al.*, 1996a), lacked intersite coherence in terms of timing, duration, and character of vegetation change. Additional fossil pollen and beetle studies from this area and the channels region (Bennett *et al.*, 2000; Hoganson, 1985; Hoganson and Ashworth, 1982; Markgraf, 1991; Villagrán, 1985; Villagrán, 1988), do not indicate climate cooling during Younger Dryas time. The co-occurrence of charcoal particles pollen assemblages of Younger Dryas-age in a few sites has been explained as direct evidence of paleoindian burning (Heusser, 1994), or decreased precipitation and high climate variability extrapolated to both sides of the Andes between 40° S and 54° S Lat. (Markgraf and Anderson, 1994).

3.3. Holocene Climate and Neoglaciation

Pollen sites between 40°-42° S (Heusser, 1982; Heusser, 1984; Villagrán, 1985; Villagrán, 1988; Villagrán, 1990) show the expansion of thermophilous Valdivian rainforest taxa, suggesting the onset of warmer/drier conditions at the onset of the Holocene, at ~9500 ^{14}C yr B.P. In contrast, Ashworth *et al.* (1991), found an increase in precipitation between 9500-5500 ^{14}C yr B.P. at the Puerto Edén site (49° S), suggesting a southward shift of the westerly storm tracks. An increase in humidity is recorded at ~5000 ^{14}C yr B.P. in pollen sites located between 40°-43° 30' S (Heusser, 1982; Heusser, 1984; Heusser *et al.*, 1995; Villagrán, 1985; Villagrán, 1988), whereas southern regions experienced a decline in precipitation (Ashworth *et al.*, 1991). The overall pattern suggests temperatures above modern values during the early Holocene, followed by a cooling event during the middle Holocene, along with a southward migration of the westerly storm tracks starting at ~9500 ^{14}C yr B.P., and a northward migration at ~5000 ^{14}C yr B.P. Superimposed on these major trends are a series of partially replicated climate signals at ~8300, ~7000-6000, and ~2500-3000 ^{14}C yr B.P. However, the precise timing and geographic extent of these events remain poorly resolved.

Glacial geological studies near the Patagonian icefields have documented glacier advances culminating at ~4500 ^{14}C yr B.P. (Mercer, 1976; Porter, 2000), ~3600 ^{14}C yr B.P. (Mercer 1976), ~2300 ^{14}C yr B.P. (Pearson I advance), ~1400 ^{14}C yr B.P. (Pearson II advance) (Aniya, 1995; Mercer, 1976), and several fluctuations since ~400 ^{14}C yr B.P., related to the Little Ice Age period (Aniya, 1995; Marden and Clapperton, 1995).

3.4. Paleoclimate Changes During and Since the LGM: A Working Hypothesis for the Southern Mid-latitudes

Ongoing research in the Región de los Lagos of southern Chile, and Lago Mascardi (Ariztegui *et al.*, 1997), in the Andes of southern Argentina, has produced high-resolution paleoclimate records spanning the LGM, the last termination, and the Holocene. These studies suggest that paleoclimate change throughout the last glacial-interglacial transition is more complex than has been previously recognized, highlighting the need for continuous, precisely dated, high-resolution studies from the icefields region to test the geographic significance of the emerging paleoclimate patterns. These results can be summarized as follows:

1. Glacial maximum conditions persisted between 29,400-14,600 ^{14}C yr B.P. At least four advances occurred within this interval, with ages of 29,400, 26,800, 22,400, and ~14,600 ^{14}C yr B.P. (Denton *et al.*, 1999a; 1999b). The climate during this interval was 6-7 °C colder and hyperhumid, with precipitation about twice the modern values. Pollen records reveal that the last pulse of extreme glacial conditions occurred between 15,800-14,600 ^{14}C yr B.P. (Heusser *et al.*, 1999; Moreno *et al.*, 1999).

2. A rapid expansion of *Nothofagus* ensued, coincident with the collapse of Andean ice lobes shortly after 14,600 ^{14}C yr B.P. North Patagonian rainforest taxa expanded abruptly at 14,200 ^{14}C yr B.P., followed by the establishment of closed-canopy rainforests at ~13,000 ^{14}C yr B.P. (Heusser *et al.*, 1999; Moreno *et al.*, 1999). These events suggest warming pulses at 14,600, 14,200, and ~13,000 ^{14}C yr B.P. that brought full-glacial to near-modern conditions in ~1600 ^{14}C years.

3. Hyperhumid conditions prevailed in the Región de los Lagos between 20,000-13,000 ^{14}C yr B.P., consistent with a northward shift and/or intensification of the westerlies' storm tracks. An important transition from a hyperhumid regime to conditions similar to our modern climate took place at 13,000 ^{14}C yr B.P., suggesting that the westerlies may have shifted southward by that time, to their "interglacial" position (Moreno *et al.*, 1999).

4. Conditions approaching modern climate prevailed between ~13,000-12,200 ^{14}C yr B.P. followed by cooling events at ~12,200 and ~11,400 ^{14}C yr B.P., and then subsequent warming at 9.8 ^{14}C yr B.P. (Moreno, 2000; Moreno *et al.*, 1999; 2001).

5. A warming event led to the expansion of Valdivian rainforest taxa at 41° S between 9200-7000 ^{14}C yr B.P. The early part of this period, between 9200-8500 ^{14}C yr B.P., was characterized by precipitation regimes below modern levels, followed by a rise in rainfall between 8500-7000 ^{14}C yr B.P. (Moreno, unpublished data).

6. Cooling events at 7000 ^{14}C yr B.P. and 5000 ^{14}C yr B.P., led to the expansion of conifers characteristic of North Patagonian rainforests (Moreno, unpublished data).

7. A period of high variability in temperature and precipitation prevailed between 5000-3000 ^{14}C yr B.P. at 41° S (Moreno, unpublished data), coincident with the onset of El Niño-Southern Oscillation, as suggested by paleoclimate records from Perú and Ecuador (Rodbell *et al.*, 1999; Sandweiss *et al.*, 1996; 1997). Several glaciers in the Patagonian icefields reached multiple Neoglacial maxima during this period of overall cooler-than-present climate, and highly fluctuating conditions.

8. A rise in temperature and reduction of precipitation between 3000-1700 ^{14}C yr B.P. led to a decline in North Patagonian conifers. A subsequent increase in precipitation at 1700 ^{14}C yr B.P., led to the establishment of modern climate in Valle Longitudinal of Región de los Lagos. (Moreno, unpublished data).

The timing of paleoclimate events outlined above correlate with several events recorded elsewhere in South America, Antarctica, and the Northern Hemisphere. Some of these events are:

   i.   the Oldest Dryas-age warming in the foraminifera record of core V29-191 in the North Atlantic Ocean (Lagerklint and Wright, 1999), the onset of deglaciation in core TR163-31B from the eastern equatorial Pacific (Shackleton, 1987), and prominent ice recession in the Swiss Alps (Schlüchter, 1988), the Laurentide and Scandinavian ice sheets from their maximum position (Lundquist and Saarnisto, 1995).

   ii.   Bölling-age warming in Europe and Greenland, and rejuvenation of North Atlantic Deep Water formation (Lehman and Keigwin, 1992; Lowe et al., 1994; Walker, 1995).

   iii. general reversal in the warming trend during late glacial time (12,000-9800 ^{14}C yr B.P.).

   iv. onset of the Holocene warming at 9800 ^{14}C yr B.P.

These results indicate a millennial-scale sequence of climate changes throughout the last termination, falling in the range of climate variability documented in high-resolution records from the North Atlantic region and polar ice cores. This cyclicity is also observed in a pollen record from Región de los Lagos throughout the last 13,000 ^{14}C years (Moreno, unpublished data), and a marine sediment core from the continental slope of the Región de los Lagos (41° S) during the last 7000 ^{14}C years (Lamy et al., 2001). The results outlined above suggest that the major climate transitions in northern Patagonia during the last 29,400 ^{14}C yr B.P. were in phase with the Northern Hemisphere, thus ruling out hypotheses stressing the role of insolation mechanisms and thermohaline switches in the North Atlantic Ocean (Broecker, 1997). Detailed studies on the tempo and mode of climate fluctuations during the last glacial-interglacial cycle are needed to test these ideas.

If the findings and interpretation of paleoclimate records from the southern mid-latitudes (and western Patagonia in particular) are correct, then they (re)open Broecker and Denton's (1989) original question: What drives ice ages? Although advances in Quaternary paleoclimatology have produced a considerable amount of new data and understanding of mechanisms related to sensitive components of the climate system, the emerging complexity is still unresolved. The absence of major changes in North Atlantic thermohaline circulation (the so-called "conveyor belt") during some Dansgaard/Oeschger events, and throughout the Holocene (Bond et al., 1999), undermines the conveyor belt paradigm as the prime mover of millennial-scale climate variability, and calls for another mechanism capable of generating changes in a substantial amount of heat in the atmosphere. According to Cane and Clement (1999) and Pierrehumbert (2000), the tropical Pacific Ocean coupled ocean-atmosphere system emerges as a likely scenario for the initiation of global and simultaneous climate changes, in a manner analogous to the modern functioning of El Niño/Southern Oscillation.

The tropical Pacific Ocean has often been a neglected or underestimated component of the climate system, mostly due to the long-standing concept of sea surface temperature (SST) stability over the tropical Pacific Ocean during the LGM, as postulated by CLIMAP (1981). Subsequent stratigraphic and modeling studies have shown otherwise (see discussion in Hostetler and Mix (1999) and references therein). The possible role of the tropical Pacific as a major mover of the climate system, with direct consequences in western Patagonia, stems from its capability of driving:

i. regional-scale changes in SSTs via changes in upwelling intensity and the Humboldt current (Sandweiss *et al.*, 1996).
ii. hemispheric-scale influences by affecting the strength and position of westerly storm tracks via changes in meridional atmospheric pressure gradients (Trenberth, 1991).
iii. global-scale effects via changes in centers of atmospheric convection (Cane and Clement, 1999), and overall water vapor content of the tropical atmosphere (Broecker, 1997).

Similarities and differences exist in terms of the timing, direction, magnitude, and rates of climate change among individual records from western Patagonia, along with important contrasts with ice-core records from interior Antarctica. It is possible that climate change was not uniform in the southern mid-latitudes during the last glacial-interglacial transition (Bennett *et al.*, 2000; Moreno *et al.*, 2001). Understanding the climate evolution of western Patagonia requires additional high-quality paleoclimate records, including ice-core records, along with sedimentary (glacial geology, limnogeology, and palynology/paleoecology) studies in the Patagonian Icefields region.

4. REFERENCES

Alley, R. B., Meese, D. A., Shuman, A. J., Gow, A. J., Taylor, K. C., Grootes, P. M., White, J. W. C., Ram, M., Waddington, E. D., Mayewski, P. A., and Zielinski, G. A., 1993, Abrupt accumulation increase at the Younger Dryas termination in the GISP2 ice core, *Nature*, **362**:527-529.

Andrews, J. T., 1998, Abrupt changes (Heinrich events) in late Quaternary North Atlantic marine environments: a history and review of data and concepts, *Journal of Quaternary Science*, **13**:3-16.

Aniya, M., 1995, Holocene glacial chronology in Patagonia: Tyndall and Upsala Glaciers, *Arctic and Alpine Research*, **27**:311-322.

Ariztegui, D., Bianchi, M. M., Masaferro, J., Lafargue, E., and Niessen, F., 1997, Interhemispheric synchrony of Late-glacial climatic instability as recorded in proglacial Lake Mascardi, Argentina, *Journal of Quaternary Science*, **12**:333-338.

Ashworth, A. C., Markgraf, V., and Villagrán, C., 1991, Late Quaternary climatic history of the Chilean Channels based on fossil pollen and beetle analysis, with an analysis of the modern vegetation and pollen rain, *Journal of Quaternary Science*, **6**:279-291.

Atkinson, T. C., Briffa, K. R., and Coope, G. R., 1987, Seasonal temperatures in Britain during the past 22,000 years, reconstructed using beetle remains, *Nature*, **325**:587-592.

Bard, E., Arnold, M., Maurice, P., Duprat, J., Moyes, J., and Duplessy, J. C., 1987, Retreat velocity of the North Atlantic polar front during the last deglaciation determined by ^{14}C accelerator mass spectrometry, *Nature*, **328**:791-794.

Behl, R. J., and Kennett, J. P., 1996, Brief interstadial events in the Santa Barbara basin, NE Pacific, during the past 60 kyr., *Nature*, **379**:243-246.

Bennett, K. D., Haberle, S. G., and Lumley, S. H., 2000, The last glacial-Holocene transition in Southern Chile, *Science*, **290**:325-328.

Björk, S., Walker, M. J. C., Cwynar, L. C., Johnsen, S., Knudsen, K.-L., Lowe, J. J., Wohlfarth, B., and INTIMATE members, 1998, An event stratigraphy for the Last Termination in the North Atlantic region based on the Greenland ice-core record: a proposal by the INTIMATE group, *Journal of Quaternary Science*, **13**:281-292.

Bond, G., Heinrich, H., Broecker, W. S., Labeyrie, L., McManus, J., Andrews, J., Huon, S., Jantschik, R., Clasen, S., Simet, C., Tedesco, K., Klas, M., Bonani, G., and Ivy, S., 1992, Evidence for massive discharges of icebergs into the North Atlantic during the last glacial period, *Nature*, **360**:245-249.

Bond, G., and Lotti, R., 1995, Iceberg discharges into the North Atlantic on millennial time scales during the last glaciation, *Science*, **267**:1005-1010.

Bond, G. C., Showers, W., Elliot, M., Evans, M., Lotti, R., Hajdas, I., Bonani, G., and Johnsen, S., 1999, The North Atlantic's 1-2 kyr climate rhythm: relation to Heinrich events, Dansgaard/Oeschger cycles and the Little Ice Age, in: *Mechanisms of Global Climate Change at Millennial Time Scales*, P. U. Clark, R. S. Webb, and L. D. Keigwin, eds., Geophysical Monograph series, Washington, DC, pp. 35-58.

Broecker, W. S., 1991, The great global conveyor, *Oceanography,* 4:79-89.

Broecker, W. S., 1994, Massive iceberg discharges as triggers for global climate change, *Nature,* 972:421-424.

Broecker, W. S., 1995, Chaotic climate, *Scientific American,* 273:62-68.

Broecker, W. S., 1997, Thermohaline circulation, the achilles heel of our climate system: will man made CO_2 upset the current balance? *Science,* 278:1582-1588.

Broecker, W. S., 1998, Paleocean circulation during the last deglaciation: A bipolar seesaw? *Paleoceanography,* 13:119-121.

Broecker, W. S., Bond, G., Mieczyslawa, K., Clark, E., and McManus, J., 1992, Origin of the northern Atlantic's Heinrich events, *Climate Dynamics,* 6:265-273.

Broecker, W. S., and Denton, G. H., 1989, The role of ocean-atmosphere reorganizations in glacial cycles, *Quaternary Science Reviews,* 9:305-341.

Cane, M., and Clement, A., 1999, A role for the Tropical Pacific coupled ocean-atmosphere system on Milankovitch and millennial timescales. Part II: Global impacts, in: *Mechanisms of Global Climate Change at Millennial Time Scales*, P. U. Clark, R. S. Webb, and L. D. Keigwin, eds., Geophysical Monograph series, Washington, DC, pp. 373-384.

Clapperton, C. M., 1991a, Glacier fluctuations of the last glacial-interglacial cycle in the Andes of South America, *Bamberger Geographische Schriften,* Bd. 11:183-207.

Clapperton, C. M., 1991b, Influence of tectonics on the extent of Quaternary glaciation in the Andes. *Bulletin IG-USP, Pub. Esp.,* 8:89-108.

CLIMAP Project Members, 1981, *Seasonal Reconstruction of the Earth's Surface at the Last Glacial Maximum*, Geological Society of America Map and Chart Series, Boulder, Colorado.

Dansgaard, W., White, J. W. C., and Johnsen. S. J., 1989, The abrupt termination of the Younger Dryas climate event, *Nature,* 339:532-533.

Denton, G. H., 2000, Does an asymmetric thermohaline-ice-sheet oscillator drive 100,000-yr glacial cycles? *Journal of Quaternary Science,* 15:301-318.

Denton, G. H., Lowell, T. V., Moreno, P. I., Andersen, B. G., and Schlüchter, C., 1999a, Geomorphology, stratigraphy, and radiocarbon chronology of Llanquihue drift in the area of the southern Lake District, Seno Reloncaví, and Isla Grande de Chiloé, Chile, *Geografiska Annaler,* 81 A:167-229.

Denton, G. H., Lowell, T. V., Moreno, P. I., Andersen, B. G., and Schlüchter, C., 1999b, Interhemispheric linkage of paleoclimate during the last glaciation, *Geografiska Annaler,* 81 A:107-153.

Grootes, P. M., Stuiver, M., White, J. W. C., Johnsen, S., and Jouzel, J., 1993, Comparison of oxygen isotope records from the GISP2 and GRIP Greenland ice cores, *Nature,* 366:552-544.

Hays, J. D., Imbrie, J., and Shackleton, N. J., 1976, Variations in the Earth's orbit: pacemaker of the ice ages, *Science,* 194:1121-1132.

Heinrich, H., 1988, Origin and consequences of cyclic ice rafting in the northeast Atlantic Ocean during the past 130,000 years, *Quaternary Research,* 29:142-152.

Heusser, C. J., 1966, Late-Pleistocene pollen diagrams from the province of Llanquihue, southern Chile, *Proceedings of the American Philosophical Society,* 110:269-305.

Heusser, C. J., 1974, Vegetation and climate of the southern Chilean Lake District during and since the last interglaciation, *Quaternary Research,* 4:190-315.

Heusser, C. J., 1981, Palynology of the last interglacial-glacial cycle in mid-latitudes of southern Chile, *Quaternary Research,* 16:293-321.

Heusser, C. J., 1982, Palynology of cushion bogs of the Cordillera Pelada, province of Valdivia, Chile, *Quaternary Research,* 17:71-92.

Heusser, C. J., 1984, Late-glacial-Holocene climate of the Lake District of Chile, *Quaternary Research,* 22:77-90.

Heusser, C. J., 1989, Southern westerlies during the last glacial maximum, *Quaternary Research,* 31:423-425.

Heusser, C. J., 1990, Chilotan piedmont glacier in the Southern Andes during the Last Glacial Maximum, *Revista Geológica de Chile,* 17:3-18.

Heusser, C. J., 1993, Late-glacial of southern South America. *Quaternary Science Reviews,* 12:345-350.

Heusser, C. J., 1994, Paleoindians and fire during the late Quaternary in southern South America, *Revista Chilena de Historia Natural,* 67:435-442.

Heusser, C. J., Denton, G. H., Hauser, A., Andersen, B. G., and Lowell, T. V., 1995, Quaternary pollen records from the Archipiélago de Chiloé in the context of glaciation and climate, *Revista Geológica de Chile,* **22**:25-46.

Heusser, C. J., and Flint, R. F., 1977, Quaternary glaciations and environments of northern Isla Grande de Chiloé, Chile, *Geology,* **5**:305-308.

Heusser, C. J., Heusser, L. E., and Lowell, T. V., 1999, Paleoecology of the southern Chilean Lake District-Isla Grande de Chiloé during Middle-Late Llanquihue glaciation and deglaciation, *Geografiska Annaler,* **81** A:231-284.

Heusser, C. J., Lowell, T. V., Heusser, L. E., Hauser, A., Andersen, B. G., and Denton, G. H., 1996a, Fullglacial-late-glacial palaeoclimate of the Southern Andes: Evidence from pollen, beetle, and glacial records, *Journal of Quaternary Science,* 11:173-184.

Heusser, C. J., Lowell, T. V., Heusser, L. E., Hauser, A., Andersen, B. G., and Denton, G. H., 1996b, Vegetation dynamics and paleoclimate during late Llanquihue glaciation in Southern Chile, *Bamberger Geographische Schriften,* Bd. **15**:201-218.

Heusser, C. J., Streeter, S. S., and Stuiver, M., 1981, Temperature and precipitation record in southern Chile extended to 43,000 yr ago, *Nature,* **294**:65-67.

Hoganson, J. W., 1985, *Late Quaternary Environmental and Climatic History of the Southern Chilean Lake Region interpreted from Coleopteran (beetle) Assemblages.* Ph. D. thesis, University of North Dakota, U.S.A.

Hoganson, J. W., and Ashworth, A. C., 1982, The late-glacial climate of the Chilean Lake region implied by fossil beetles, B. Mamet and M. J. Copeland, eds., *Proceedings North American Paleontological Convention,* 3:251-256.

Hoganson, J. W., and Ashworth, A. C., 1992, Fossil beetle evidence for climatic change 18,000-10,000 years B.P. in south-central Chile, *Quaternary Research,* **37**:101-116.

Hostetler, S. W., and Mix, A. C. 1999, Reassessment of ice-age cooling of the tropical ocean and atmosphere, *Nature,* **399**, 673-676.

Hughen, K. A., Overpeck, J. T., Lehman, S. J., Kashgarian, M., Southon, J., Peterson, L. C., Alley, R., and Sigman, D. M., 1998, Deglacial changes in ocean circulation from an extended radiocarbon calibration, *Nature,* **391**:65-68.

Imbrie, J., Boyle, E. A., Clemens, S. C., Duffy, A., Howard, W. R., Kukla, G., Kutzbach, J., Martinson, D. G., McIntyre, A., Mix, A. C., Molfino, B., Morley, J. J., Peterson, L. C., Pisias, N. G., Prell, W. L., Raymo, M. E., Schackleton, N. J., and Toggweiler, J. R., 1992, On the structure and origin of major glaciation cycles 1. Linear responses to Milankovitch forcing, *Paleoceanography,* **7**:701-738.

Lagerklint, I. M., and Wright, J. D., 1999, Late glacial warming prior to Heinrich event 1: The influence of ice rafting and large ice sheets on the timing of initial warming, *Geology,* **27**:1099-1102.

Lamy, F., Hebbelm, D., Rohl, U., and Wefer, G., 2001, Holocene rainfall varibility in southern Chile: a marine record of latitudinal shifts of the southern westerlies, *Earth and Planetary Science Letters,* **185**:369-382.

Laugenie, C., 1982, *La region des lacs, Chili meriodonal, recherches sur l'évolution géomorphologique d'un piédemont glaciaire quaternaire andin,* Ph. D. Thesis, l'Université de Boreaux, France (in French).

Lehman, S. J., and Keigwin, L. D., 1992, Sudden changes in North Atlantic circulation during the last deglaciation, *Nature,* **356**:757-762.

Lowe, J. J., Ammann, B., Birks, H. H., Bjørk, S., Coope, G. R., Cwynar, L., De Beaulieu, Mott, R. J., Peteet, D. M., and Walker, M. J. C., 1994, Climatic changes in areas adjacent to the North Atlantic during the last glacial-interglacial transition (14-9 ka B.P.): a contribution to IGCP-253, *Journal of Quaternary Science* **9**:185-198.

Lowell, T. V., Heusser, C. J., Andersen, B. G., Moreno, P. I., Hauser, A., Denton, G. H., Heusser, L. E., Schluchter, C., and Marchant, D, 1995, Interhemispheric correlation of Late Pleistocene glacial events, *Science,* **269**:1541-1549.

Lumley, S. H., and Switsur, R., 1993, Late Quaternary of the Taitao Peninsula, Southern Chile, *Journal of Quaternary Science,* **8**:161-165.

Lundquist, J., and Saarnisto, M. 1995, Summary of project IGCP-253, *Quaternary International,* **28**:9-18.

Marden, C. J., 1997, Late-glacial fluctuations of South Patagonian Icefield, Torres del Paine National Park, southern Chile, *Quaternary International,* **38/39**:61-68.

Marden, C. J., and Clapperton, C. M., 1995, Fluctuations of the South Patagonian Ice-Field during the last glaciation and the Holocene, *Journal of Quaternary Science,* **10**:197-210.

Markgraf, V., 1989, Reply to Heusser's "Southern westerlies during the Last Glacial Maximum", *Quaternary Research,* **31**:426-432.

Markgraf, V., 1991, Younger Dryas in South America? *Boreas,* **20**:63-69.

Markgraf, V., and Anderson, L., 1994, Fire history in Patagonia: climate versus human cause, *Revista Instituto de Geología,* **1**:35-47.

Mayewski, P. A., Meeker, L. D., Whitlow, S., Twickler, M. S., Morrison, M. C., Alley, R. B., Bloomfield, P., and Taylor, K., 1993, The atmosphere during the Younger Dryas, *Science,* **261**:195-197.

Mayewski, P. A., Twickler, M. S., Whitlow, S. I., Meeker, L. D., Yang, Q., Thomas, J., Kreutz, K., Grootes, P. M., Morse, D. L., Steig, E. J., Waddington, E. D., Saltzman, E. S., Whung, P.-Y., and Taylor, K. C., 1996, Climate change during the last deglaciation in Antarctica, *Science,* **272**:1636-1638.

McCulloch, R. D., Bentley, M. J., Purves, R. S., Hulton, N. R. J., Sugden, D. E., and Clapperton, C. M., 2000, Climatic inferences from glacial and palaeoecological evidence at the last glacial termination, southern South America, *Journal of Quaternary Science,* **15**:409-417.

Mercer, J. H., 1976, Glacial history of southernmost South America, *Quaternary Research,* **6**:125-166.

Moreno, P. I., 1997, Vegetation and climate change near Lago Llanquihue in the Chilean Lake District between 20,200 and 9500 ^{14}C yr B.P., *Journal of Quaternary Science,* **12**:485-500.

Moreno, P. I., 2000, Climate, fire, and vegetation between about 13,000 and 9200 ^{14}C yr B.P., *Quaternary Research,* **54**:91-89.

Moreno, P. I., Jacobson, G. L., Andersen, B. G., Lowell, T. V., and Denton, G. H., 1999, Vegetation and climate changes during the last glacial maximum and the last termination in the Chilean Lake District: A case study from Canal de la Puntilla (41°S), *Geografiska Annaler,* **81 A**:285-311.

Moreno, P. I., Jacobson, G. L., Lowell, T. V., and Denton, G. H., 2001, Interhemispheric climate links revealed from a late-glacial cool episode in southern Chile, *Nature,* **409**:804-808.

Pierrehumbert, R. T., 2000, Climate change and the tropical Pacific: The sleeping dragon wakes, *Proceedings of the National Academy of Sciences,* **97**:1355-1358.

Porter, S. C., 1981, Pleistocene glaciation in the southern lake district of Chile, *Quaternary Research,* **16**:263-292.

Porter, S. C., 2000, Onset of neoglaciation in the Southern Hemisphere, *Journal of Quaternary Science,* **15**:395-408.

Rodbell, D. T., Seltzer, G. O., Anderson, D. M., Abbott, M. B., Enfield, D. B., and Newman, J. H., 1999, An ~15,000-year record of El Niño-driven alluviation in southwestern Ecuador, *Science,* **283**:516-520.

Sandweiss, D. H., Richardson III, J. B., Reitz, E. J., Rollins, H. B., and Maasch, K. A., 1997, Determining the beginning of El Niño: Response (to comments), *Science,* **276**:966-967.

Sandweiss, D. H., Richardson, J. B., Reitz, E. J., Rollins, H. B., and Maasch, K. A., 1996, Geoarcheological evidence from Perú for a 5000 years B.P. onset of El Niño, *Science,* **273**:1531-1533.

Schlüchter, C., 1988, The deglaciation of the Swiss Alps: A paleoclimate event with chronological problems, *Bulletin de l'Association Francaise Pour l'Etude du Quaternaire,* **2/3**:141-145.

Shackleton, N. J., 1987, Oxygen isotopes, ice volume, and sea level, *Quaternary Science Reviews,* **6**:183-190.

Shackleton, N. J., 2000, The 100,000-year Ice-Age cycle identified and found to lag temperature, carbon dioxide, and orbital eccentricity, *Science,* **289**:1897-1902.

Sowers, T., and Bender, M., 1995, Climate records covering the last deglaciation, *Science,* **269**:210-214.

Toggweiler, J. R., and Samuels, D., 1995, Effect of Drake Passage on the global thermohaline circulation, *Deep Sea Research,* **1**:477-500.

Trenberth, K. E., 1991, Storm tracks in the southern hemisphere, *Journal of Atmospheric Sciences,* **48**:2159-2178.

Villagrán, C., 1985, Análisis palinológico de los cambios vegetacionales durante el Tardiglacial y Postglacial en Chiloé, Chile, *Revista Chilena de Historia Natural,* **58**:57-69 (in Spanish).

Villagrán, C., 1988, Late Quaternary vegetation of Southern Isla Grande de Chiloé, Chile, *Quaternary Research,* **29**:294-306.

Villagrán, C., 1990, Glacial, Late-Glacial, and Post-Glacial climate and vegetation of the Isla Grande de Chiloé, southern Chile (41-44°S), *Quaternary of South America and Antarctic Peninsula,* **8**:1-15.

Walker, M. J. C., 1995, Climatic changes in Europe during the Last Glacial/Interglacial transition, *Quaternary International,* **28**:63-76.

Wenzens, G., 1999, Fluctuations of outlet and valley glaciers in the Southern Andes (Argentina) during the past 13,000 years, *Quaternary Research,* **51**:238-247.

LATE PLEISTOCENE AND HOLOCENE GLACIER FLUCTUATIONS IN THE MENDOZA ANDES, ARGENTINA

Lydia E. Espizua[1*]

1. ABSTRACT

The río Mendoza valley at 33° S latitude has been repeatedly invaded by glaciers during the Late Pleistocene. Relative-age criteria, U-series ages and thermoluminescence, permitted the glacial deposits to be separated into three mappable units, each less extensive than its predecessor. Interstadial sediments in the upper río Mendoza valley were dated in order to constrain the ages of the drifts. The moraine sequence is compared with those studied by Espizua (1998) along the río Grande valley in the southwestern Mendoza Province at 35° S latitude.

In the río Valenzuela valley located at 35° S latitude, in the upper río Grande basin, the Holocene glacier variations were recognized in the El Azufre and El Peñón valleys. A first Neoglacial advance occurred at ca. 5700-4400 yr B.P. A second Neoglacial advance occurred at ca. 2500-2200 yr B.P. A third Neoglacial advance, identified as the Little Ice Age, culminated at ca. 400 yr B.P. (ca. 1435–1660 cal AD yr).

2. INTRODUCTION

The Andes of Mendoza have a north-south orientation and extend between latitudes 33° and 37° South. At latitude 33° S, the highest peaks are generally over 5000 m and culminate in cerro Aconcagua (6959 m), the highest mountain in the Western Hemisphere, while to the south at ca. 37° S, the highest altitudes range between 3000 and 4000 m. The study area is limited on the west by the mountain range between Chile and Argentina. The Andes act as a barrier for most of the humidity coming from the Pacific Ocean westerlies, so on the eastern side of the Andes the climate is semi-arid. To the

[1] Instituto Argentino de Nivología, Glaciología y Ciencias Ambientales, Consejo Nacional de Investigaciones Científicas y Técnicas, Casilla de Correo 330 (5500), Mendoza, Argentina.

* corresponding author: lespizua@lab.cricyt.edu.ar

The Patagonian Icefields: A Unique Natural Laboratory for Environmental and Climate Change Studies.
Edited by Gino Casassa et al., Kluwer Academic /Plenum Publishers, 2002.

south, the decreasing altitude of the mountains gradually allows a larger penetration of moist westerly winds.

The mean annual precipitation is about 300-400 mm in the río Mendoza valley, and about 940 mm in the río Grande valley. The mean monthly precipitation and temperature records show a well-defined winter precipitation maximum, a temperature minimum from May to August, and a dry warm summer season.

The Quaternary record in this part of the Andes is mostly undated, so this research is an important step towards obtaining ages of the drifts and paleoclimatic information about climate changes. The major objective was to define the Quaternary glaciations in both valleys of the central Andes of Mendoza and to obtain the ages of the glacial deposits in order to constrain the ages of the drifts.

Figure 1. Map of the río Mendoza valley in the Central Argentine Andes showing the distribution of the late Pleistocene moraines, a Holocene advance (Espizua, 1993) and sampling localities. Reprinted from *Global and Planetary Change,* **22**, Espizua, L., Chronology of Late Pleistocene glacier advances in the Río Mendoza valley, Argentina, pp. 193-200, (1999) with permission from Elsevier Science.

3. CHRONOLOGY OF LATE PLEISTOCENE GLACIER ADVANCES

In the río de los Horcones - río de las Cuevas valleys, five Pleistocene drifts and one Holocene drift were differentiated by Espizua (1993) and correlated by multiple relative-age criteria, stratigraphic studies and some absolute ages. Three late Pleistocene drifts that may have been deposited during the last glaciation were named from oldest to youngest, Penitentes, Horcones and Almacenes drifts (Figure 1). During the Penitentes advance, the glacier system flowed down the Las Cuevas and the Los Horcones valleys to an altitude of 2500 m while during the subsequent Horcones advance, an independent ice stream from Los Horcones Inferior and Superior valleys occupied Los Horcones valley and extended to the terminal moraine at 2750 m. The Almacenes drift extends through the lower 2 km of Los Horcones Inferior valley at an altitude of 3250 m (Espizua 1989; 1993).

Previous studies have shown that along the río de las Cuevas valley, a travertine layer overlies Penitentes till. The travertine was dated by the U-series ages of 24,200 ± 2000 yr, 22,800 ± 3100 yr and 38,300 ± 5300 yr B.P. (Figure 1, sites 4 and 5), which are minimum ages for the underlying Penitentes till (Bengochea *et al.*, 1987; Espizua, 1993).

At Confluencia, where the río de los Horcones Inferior joins the río de los Horcones Superior, the Horcones till underlies the Almacenes till and rests, in turn, on the Penitentes till. In some exposures two or three drifts rest in superposition, the contacts marked by erosional unconformities. However, in some places the drifts are separated by sediments of non-glacial origin, non-stratified and unconsolidated, composed mainly of silt-sized particles, fine sand and clay.

A date was obtained from a composite stratigraphic profile based on exposures along the east side of the río de los Horcones Inferior valley, a tributary of the río de las Cuevas valley, which includes the Penitentes and Horcones tills separated by non-glacial sediments (silt, fine sand, and clay), interpreted as representing the Penitentes-Horcones non-glacial interval (Espizua, 1999). The fine quartz grains (4-11 μm) of these sediments were TL dated as 31,000 ± 3100 yr B.P. (Figure 1, site 1). The TL 31,000 yr age of these sediments is consistent with the U-series ages of the travertine layers overlying Penitentes till (Figure 2). All these dates, which are minimum ages for the underlying Penitentes till, imply that the Penitentes ice advanced prior to the last glacial maximum, and sometime before ca. 40,000 yr ago.

Support for this interpretation is also emerging from palynological studies. Pollen analysis in the Rincón del Atuel, at latitude 34°45' S in Mendoza Province, suggests a temperature increase between 27,000 and 24,500 yr B.P. (D'Antoni, 1980). Markgraf *et al.* (1986) suggest, on the basis of pollen and diatom analyses of radiocarbon-dated lacustrine sediments, that interglacial-type climatic conditions existed between 33,000 and 27,000 yr B.P. in the temperate Andean region at 40° S latitude. The Penitentes advance likely preceded this interval of milder climate. Otherwise, based on pollen analysis the climate was cold between 24,500 and 15,500 yr B.P. in the Rincón del Atuel (D'Antoni, 1980).

A minimum date for the Horcones till comes from the exposure on the east side of the río de los Horcones Inferior valley where the Horcones and Almacenes tills are separated by sediments of non-glacial origin (Figure 1, site 2). The fine quartz grains (4-11 μm) of these sediments have been dated by TL as 15,000 ± 2100 yr B.P. (Espizua, 1999). They are also interpreted as representing the Horcones-Almacenes non-glacial interval (Figure 2). Otherwise the travertine layer capping the Horcones outwash in the río de las Cuevas valley (Figure 1, site 3) has a U-series age of 9700 ± 5000 yr B.P. (Bengochea *et al.*, 1987). Despite the large error factor, probably owing to the sample containing a large amount of limestone detritus, the date is consistent with a latest Pleistocene age for the Horcones drift. The available dates imply that the Horcones ice advance represents the culmination advance of the last glaciation. The Almacenes till is inferred to represent a stand-still or a re-advance that occurred at ca. 14,000 or between 11,000 - 10,000 yr B.P.

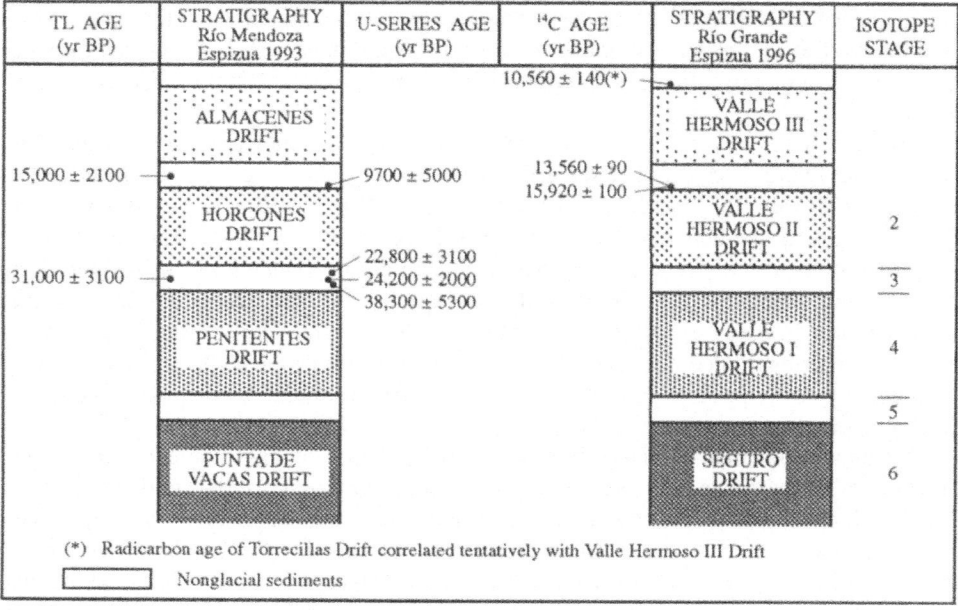

Figure 2. Generalized stratigraphy in the río Mendoza and río Grande valleys, and the TL, ^{14}C and U-series dates. Reprinted from *Global and Planetary Change*, **22**, Espizua, L., Chronology of Late Pleistocene glacier advances in the río Mendoza valley, Argentina, pp. 193-200, (1999), with permission from Elsevier Science.

Based on these studies, Penitentes advance may correlate with marine oxygen isotope stage 4, and the Horcones and the Almacenes till, may likely equate with stage 2.

The depression of the steady-state ELA for the Horcones and Penitentes advances was ca. 1000 and 1200 m, respectively (Espizua, 1993).

In the río Grande valley in the southwestern Mendoza province, between latitude 34°48' and 35°40' South, Espizua (1998) suggested, on the basis of relative age-criteria, morphology and ^{14}C dates, that during the last glaciation (valle Hermoso Drift) there were three advances, each one less extensive that its predecessor. They were named from oldest to youngest valle Hermoso I, II and III drifts.

Radiocarbon dates for the valle Hermoso II drift come from a non-basal peat sample obtained from a bog within the limit of the moraine that was dated as 13,560 ± 90 yr B.P. (Beta Analytic-71676) and a basal sample dated as 15,920 ±100 yr B.P. (Beta Analytic-92454).

The stratigraphy suggests that valle Hermoso I drift probably correlates with that of early Wisconsin. The valle Hermoso II drift is inferred to represent the culminating advance during the last glaciation, while valle Hermoso III drift is regarded as late-glacial in age (at ca. 14,000 yr or between 11,000 - 10,000 yr B.P.). In the upper río Salado valley (latitude 35° S), the Torrecillas moraine related with a peat bog behind the moraine, dated as 10,560 ± 140 yr B.P. (LATYR LP-473). This advance is considered probably equivalent in age with the valle Hermoso III re-advance (Espizua, 1998).

Based on contrast in morphology, stratigraphy, altitudinal distribution of the drifts and radiocarbon ages, it is inferred that valle Hermoso I, II and III drifts could be correlated tentatively with Penitentes, Horcones and Almacenes drift.

4. HOLOCENE GLACIER VARIATIONS

The río Valenzuela valley is located at latitude 35° S in the upper río Grande basin where two tributaries valleys named Los Baños del Azufre and El Peñón have been studied (Figure 3). The area is important from the paleoclimatic point of view, because it lies in a transition zone between the arid and semi-arid conditions that prevail in the Andes of Cuyo and the humid conditions of Patagonia to the south.

The Holocene record in this part of the Andes is mostly unknown and there have only been a few studies carried out on this subject. The purpose of this study was to differentiate the Holocene glacier variations in order to obtain the glacier chronology of this area.

5. THE AZUFRE VALLEY

Differentiation of the drifts was based mainly on relative-age criteria, morphology, stratigraphic relationships and absolute ages. The glacial deposits have been divided into four mappable units, named from oldest to youngest, Los Baños del Azufre, Pertucio, El Macho and El Fierro drifts (Figure 3).

The lateral-terminal moraine Los Baños del Azufre is well defined and it lies at 2460 m. A concentration of large boulders of volcanic rock marks the limit of the terminal moraine. In the south sector of the moraine, a basal peat bog sample was collected and yielded an age of 5710 ± 70 yr B.P. (LP 717). This moraine is related to the Los Baños del Azufre outwash terrace that is covered by a volcanic ash layer between 0.80 and 1.70 meters thick. According to Naranjo and Haller (1997) the volcanic deposits correspond to the Pomez de los Baños. A horizon of organic material lying under the volcanic deposit was dated as 1400 ± 80 yr and 1050 ± 90 yr B.P. Similar data, 1580 ± 50 yr B.P., was obtained from a sample (LP 508), collected from an organic horizon of 0.20 m thick lying on the terrace on the right side of the Arroyo Azufre.

The data of 5710 yr suggests that this advance could correspond to the start of the First Neoglaciation.

The Pertucio terminal moraine ended at 2470 m, and lies very close to the limit of the Los Baños del Azufre drift. A radiocarbon date of 4770 ± 70 yr B.P. (LP 622) was obtained from a peat bog inside the limit of the moraine in contact with the till. The First

Neoglacial advance probably culminated at ca. 4700 yr B.P. This advance could be correlated with the Confluencia drift (Figure 1) in the Mendoza valley (Espizua, 1993).

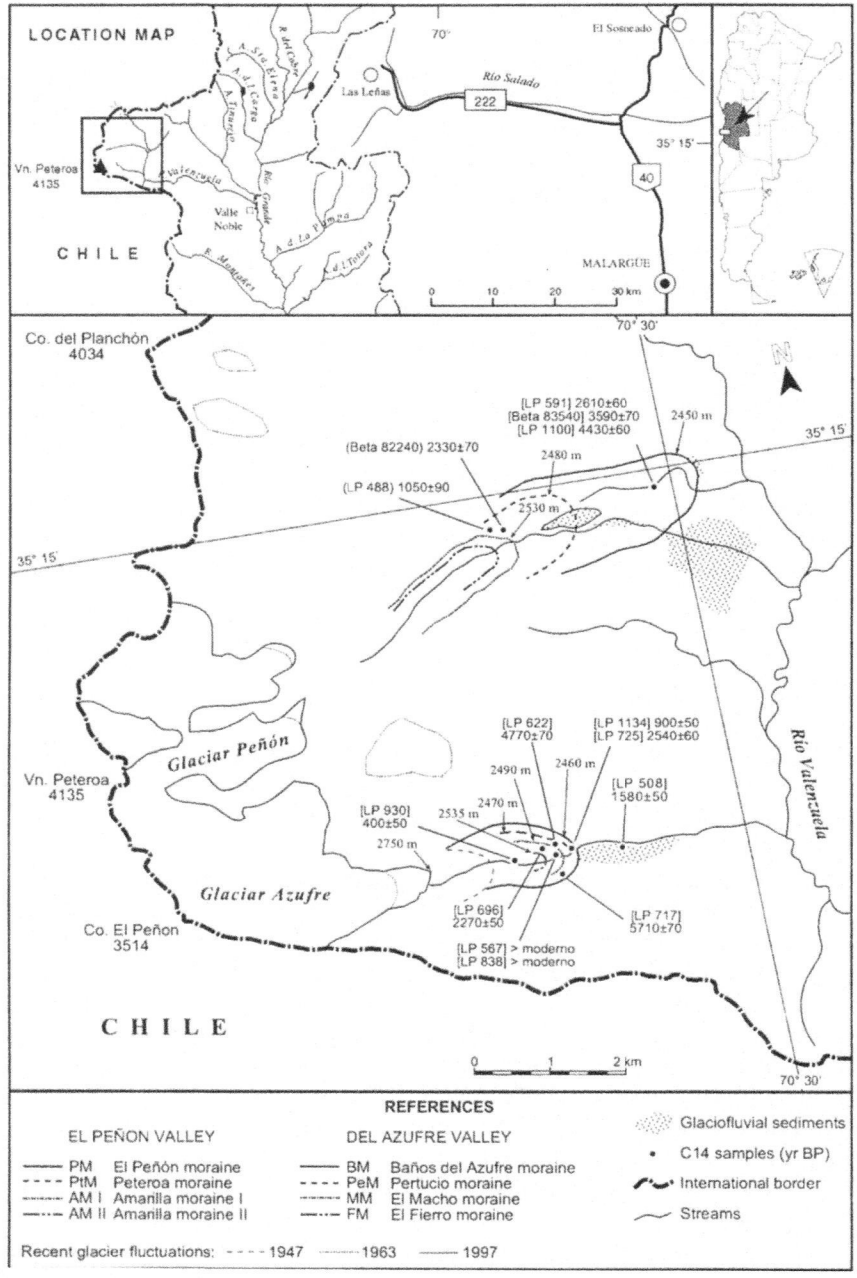

Figure 3. Map of the río Valenzuela valley showing the Holocene glacier fluctuations.

Similar ages for this advance were found by many authors through palynological and glaciological studies in the Andes of southern Chile and Argentina (Mercer, 1976; Röthlisberger, 1986; Aniya and Sato, 1995; Aniya, 1995; Clapperton and Sugden, 1988; Markgraf and Bradbury, 1982; Stingl and Garleff, 1985).

D' Antoni (1980) suggests, based on pollen analysis in the cave Gruta del Indio in the río Atuel, at lat. 35° S, an increase in temperature and aridity between 9000 and 5000 yr. B.P. Between 5000 and 2000 yr. B.P., he suggests an increase in precipitation, presumably in the form of winter rains and also implying lower temperatures. This climatic configuration lasted until 2000 yr. B.P. when modern climatic conditions became established, with summer rain precipitation in the lowlands and more favorable temperatures in the highlands. Markgraf (1983) compared the pollen sequence analyzed by D' Antoni (1980) with two high elevation pollen records from peat bogs in Salina 2 in the upper Uspallata valley (Mendoza basin) and Salado (río Salado basin), and established the basic paleoenvironmental chronology from the desert area at latitude 30° to 34° S.

The second Neoglaciation is indicated by the El Macho moraine which ended at 2490 m. The moraine is partially eroded, although the outwash terrace is well defined. A radiocarbon age from basal peat on the moraine was dated as 2270 ± 50 yr B.P. (LP 696) and ages of 2540 ± 60 (LP 725) and 900 ± 50 yr B.P. (LP 1134) proceeded from the related outwash terrace. The basal peat samples were obtained on the outwash near the hydrothermal water bath named El Pertucio.

Grosjean et al., (1998) suggest a glacier advance younger than 2630 ± 45 yr B.P. in the western part of the Arid Andes at 29° S, northern Chile, based on radiocarbon dates and relative ages of soil development on moraines.

The date of the third Neoglaciation, around 400 ± 50 yr B.P., (cal AD yr 1435–1660) (LP 930) was obtained from a basal peat inside the limit of the moraine. The El Fierro moraine which extends to an altitude of 2535 m, and the related outwash terrace, are poorly developed. Two samples of peat on the outwash terrace were dated (> modern; LP 838 and 567). The freshness of the drift, together with the proximity to the active glacier, suggest that this advance could correspond to the Little Ice Age.

This date is consistent with the glacial chronologies proposed for Patagonia east of the Andes by Villalba et al. (1990); Rabassa et al. (1984) and Mercer (1976; 1982).

6. THE EL PEÑÓN VALLEY

In the El Peñón valley four drifts were differentiated. From older to youngest they were named: El Peñón, Peteroa and Amarilla I and II drifts (Figure 3). The outermost moraine named El Peñón is well preserved and the terminal moraine lies at an altitude of 2450 m. The terminal moraine grades into an outwash terrace that is partially covered by pyroclastic deposits. The unweathered till is grayish-brown (10YR, according to Munsell soil color charts) and contains angular to subangular boulders of volcanic rocks on the moraine crest. The boulders present a poor development of varnish.

The pyroclastic deposits of the El Planchón volcano extend to the east side of the volcano. According to Naranjo and Haller (1997) the axis of the Oleada Piroclástica Valenzuela, a gray volcanic deposit, trends to the south-east. It crops out for about 25 km along the Valenzuela valley from its eruptive center. The deposit is about 4 m in depth and presents layers with internal cross and graded-bedding sedimentary structure. Naranjo and Haller have dated two samples interbedded with the ash deposits by Accelerator Mass Spectrometry. The ages are 7030 ± 70 and 7020 ± 60 yr B.P. (Beta Analytic lab). These dates imply that the Oleada Piroclástica Pumícea is older than the El Peñón drift.

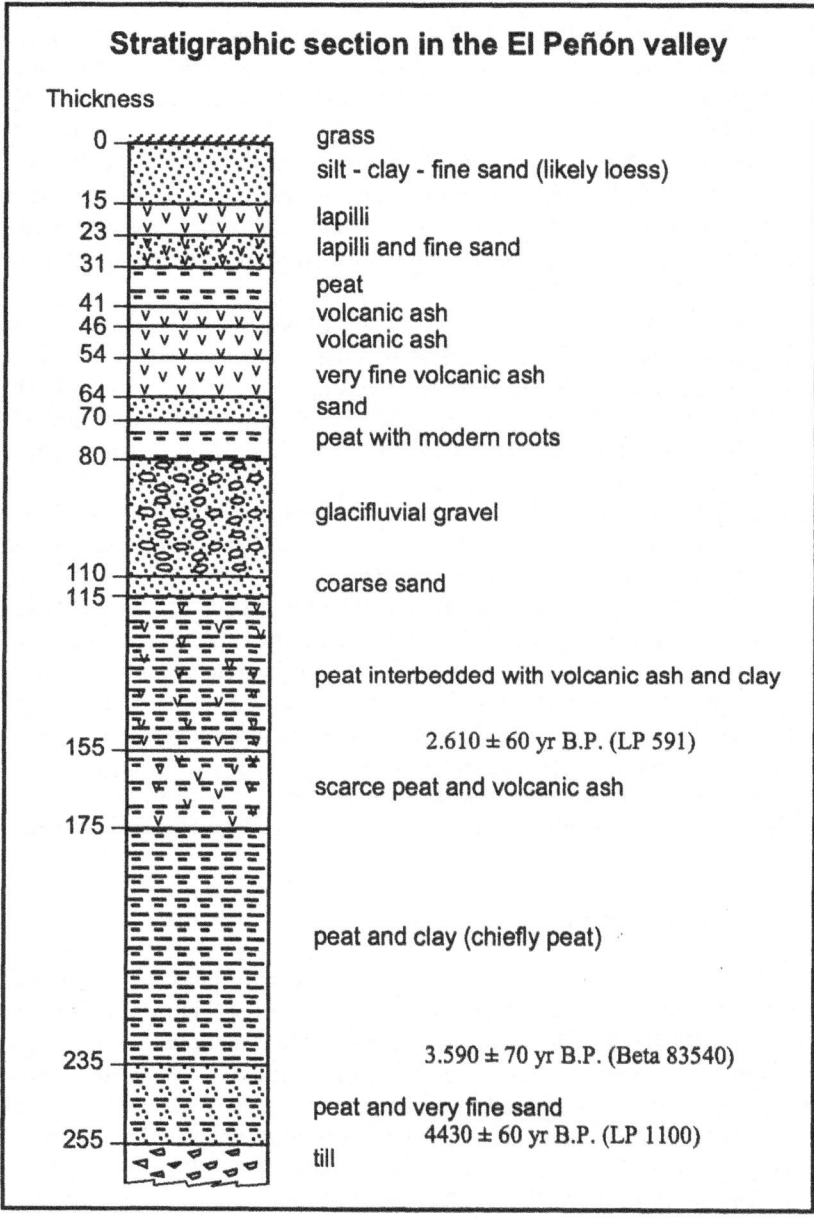

Figure 4. Stratigraphic profile in the río El Peñón valley, showing the location of the samples dated by [14]C. Thickness units correspond to cm.

A natural profile was studied about 200 m north of the Campamento of Gendarmería on the left side of a stream channel. The profile of about 2.20 m thickness presented peat, clay and glaciofluvial gravel horizons (Figure 4). At the base of the exposure a layer of organic material lying on the El Peñón drift was dated as 4430 ± 60 yr B.P. (LP1100). A peat horizon that overlies it, was dated as 3590 ± 70 yr B.P. (Beta 83540). To the top of the profile another peat horizon was dated as 2610 ± 60 yr B.P. (LP 591). Collectively, these dates imply that the El Peñón advance occurred during the first Neoglaciation, ca. 4400 yr B.P.

The morphology of the moraine assigned to the second Neoglacial advance (Peteroa moraine) is not well defined, it is partially eroded and covered on the valley floor by peat. The terminal moraine ended probably at 2480 m. Inside the limit of the Peteroa moraine, two basal samples were dated to be 2330 ± 70 yr B.P. (Beta 82240) and 1050 ± 90 yr B.P. (LP 488).

The third Neoglacial advance is indicated by the Amarilla I and II moraines, which are named thus for their yellow color. The Amarilla I moraine is well preserved and ended at 2530 m. Lateral yellow moraines built along both margins of the valley near to the glacier are well preserved. No absolute ages have been obtained up to now for these moraines, but the freshness of the drift, together with the proximity to the active glacier, suggest that this advance may correspond to the Little Ice Age (third Neoglacial advance).

Based on relative-age chronology, morphology, stratigraphic relationships and ^{14}C dates, the Los Baños del Azufre and the El Pertucio moraines, assigned to the first Neoglacial advance, could be correlated with the El Peñón moraine; El Macho moraine may likely correlate with the Peteroa moraine. The El Fierro and the Amarilla I and II moraines were deposited presumably during the Little Ice Age.

In southern South America, Mercer (1982) identified three Neoglacial advances at ca. 4500-4000 yr B.P.; 2700-2000 yr B.P., and during recent centuries. The three Neoglacial advances recognized in the río Valenzuela valley area support Mercer's model for the Andes. However much more evidence and studies are required in the area. It is interesting that the Little Ice Age, which was first noticed in Europe from historical records, is found here and thus was presumably global in extent.

7. RECENT GLACIER FLUCTUATIONS

Observations of the present glacier fluctuations in the río Valenzuela valley were made based on historic information, aerial photographs of 1963 and a satellite image of 1997. The Azufre Glacier flows to the east and the front reached an altitude of 2750 m, through field measurements using an altimeter. Between 1947 (personal communications, Vialidad Provincial) and 1963, the El Azufre Glacier has receded 1285 m, and from 1963 to 1997, the glacier advanced 430 m. Some small glaciers and snow patches that appear in the 1963 aerial photograph had disappeared in 1997.

The El Peñón Glacier flows to the south-east and presents two tongues. The south front advanced 325 m, and the north one, 120 m between 1963 and 1997. Some small glaciers that were present on the aerial photo of 1963 have disappeared as the 1997 image shows.

In general the glaciers in the central Andes of Mendoza, underwent a general retreat from the beginning of last century (Espizua, 1986). An advance of some glacier fronts was observed between 1963 and 1997, but some areas that were covered by ice on the 1963 aerial photos, are at the present time free of ice in the study area.

8. CONCLUSIONS

The valle Hermoso drift II in the río Grande basin in the south western Mendoza province, has a basal ^{14}C age of 15,920 ± 100 yr B.P. which is a minimum age for the underlying till. The available dates imply that the valle Hermoso II till represents the culmination advance of the last glaciation.

Based on previous chronostratigraphic studies in the río Mendoza valley, the Penitentes advance may correlate with marine oxygen isotope stage 4. The Horcones and Almacenes tills may likely equate with stage 2. In the río Grande valley, the valle Hermoso I, II and III drifts could be correlated tentatively with Penitentes, Horcones and Almacenes drifts.

In the río Valenzuela valley located in the upper río Grande basin, the Holocene glacier variations were differentiated and some radiocarbon ages were obtained in the El Azufre and El Peñón valleys. A first Neoglacial advance occurred at ca. 5700-4400 yr B.P. and it could be tentatively correlated with the Confluencia drift in the río Mendoza basin. A second Neoglacial advance occurred at ca. 2500-2000 yr B.P. A third Neoglacial advance, the Little Ice Age, culminated at ca. 400 yr B.P. (ca. 1435-1660 cal AD yr). The three Neoglacial advances recognized in the río Valenzuela valley support Mercer's model for the south of Chile and Argentina Andes.

Based on historic information, aerial photographs and a satellite image of 1997, the El Azufre Glacier has receded between 1945 and 1963. The El Azufre and Peñón Glaciers advanced between 1963 and 1997.

9. ACKNOWLEDGMENTS

I thank C. J. Aguado and especially R. Bottero and H. Videla for their generous help in the field. The figures were drawn by R. Bottero. Funding was provided by the Universidad Nacional de Cuyo.

10. REFERENCES

Aniya, M., 1995, Holocene glacial chronology in Patagonia: Tyndall and Upsala Glaciers, *Arctic and Alpine Research*, **27**(4):311-322.

Aniya, M., and Sato, H., 1995, Holocene glacial chronology of Upsala Glacier at Península Herminita, Southern Patagonia Icefield. Glacier Research in Patagonia. *Bulletin of Glacier Research* **13**:83-96.

Bengochea, L. E., Porter, S. C., and Schwarcz, H. P., 1987, Pleistocene glaciation across the high Andes of Chile and Argentina, in: Abstracts of the International Union of Quaternary Research INQUA, XIIth International Congress, Ottawa, Canada.

Clapperton, C. M., and Sugden, D. E., 1988, Holocene glacier fluctuations in South America and Antarctica, *Quaternary Science Reviews*, 7:185-198.

D'Antoni, H. L., 1980, Los últimos 30 mil años en el sur de Mendoza, Argentina, III Coloquio Sobre Paleobotánica y Palinología, Memorias del Instituto Nacional de Antropología e Historia, pp. 83-102 (in Spanish).

Espizua, L. E., 1986, Fluctuations of the río del Plomo Glaciers, *Geografiska Annaler*, **68A**(4), 317-327.

Espizua, L. E., 1989, *Glaciaciones Pleistocénicas en la Quebrada de los Horcones y Río de las Cuevas, Mendoza, República Argentina*, Doctoral Thesis, Universidad Nacional de San Juan, San Juan, Argentina, pp. 226 (in Spanish).

Espizua, L. E., 1993, Quaternary glaciations in the río Mendoza valley, Argentine Andes, *Quaternary Research*, **40**:150-162.

Espizua, L. E., 1998, Secuencia glacial del Pleistoceno Tardío en el valle del río Grande, Mendoza, Argentina, *Bamberger Geographische Schriften*, Bd. **15S**:85-99, Bamberg, (in Spanish).

Espizua, L. E., 1999, Chronology of Late Pleistocene glacier advances in the río Mendoza Valley, Argentina, *Global and Planetary Change*, **22**:193-200.

Grosjean, M., Geyh, M. A., Messerli, B., Schreier, H., and Veit, H., 1998, A late-Holocene (<2600 B.P.) glacial advance in the south-central Andes (29°S), northern Chile, *The Holocene,* **8**(4):473-479.

Markgraf, V., and Bradbury, P., 1982, Holocene climatic history of South America, *Striae*, **16**:40-45.

Markgraf, V., 1983, Late and postglacial vegetational and paleoclimatic changes in subantarctic, temperate and arid environments in Argentina, *Palynology*, **7**:43-70.

Markgraf, V., Bradbury, J. P., and Fernandez, J., 1986, Bajada de Rahue, Province of Neuquén, Argentina: an interstadial deposit in northern Patagonia, *Palaeogeography, Palaeoclimatology, Palaeoecology*, **56**:251-258.

Mercer, J. H., 1976, Glacial history of southernmost South America, *Quaternary Research*, **6**:125-166.

Mercer, J. H., 1982, Holocene glacier variations in southern South America, *Striae,* **18**:35-40.

Naranjo, J. A., and Haller, M., 1997, Actividad explosiva postglacial del complejo volcánico Planchón-Peteroa, 35°15' S, Universidad Católica del Norte. Departamento de Ciencias Geológicas, VIII Congreso Geológico Chileno, Actas, **I**(2):357-360 (in Spanish).

Rabassa, J., Brandani, A. A., Boninsegna, J. A., and Cobos, D. R., 1984, Cronología de la Pequeña Edad de Hielo en los glaciares Río Manso y Castaño Overo, cerro Tronador, Provincia de Río Negro, IX Congreso Geológico Argentino, Actas, **3**:624-639 (in Spanish).

Röthlisberger, F., 1986, 10,000 *Jahre Gletschergeschichte der Erde*, Verlag Sauerländer Aarau. Frankfurt am Main, Salzburg, pp. 187-416 (in German).

Stingl, H., and Garleff, K., 1985, Glacier variations and climate of the Late Quaternary in the subtropical and mid-latitude Andes of Argentina, *Zeitscrift für Gletscherkunde und Glazialgeologie*, Bd. **21S**:225-228.

Villalba, R., Leiva, J. C., Rubulls, S., Suarez, J., and Lenzano, L., 1990, Climate, tree-ring, and glacial fluctuations in the Río Frías valley, Río Negro, Argentina, *Arctic and Alpine Research*, **22**(3):215-232.

CURRENT KNOWLEDGE OF THE SOUTHERN PATAGONIA ICEFIELD

Gino Casassa[1,2*], Andrés Rivera[1,3], Masamu Aniya[4], and Renji Naruse[5]

1. ABSTRACT

We present here a review of the current glaciological knowledge of the Southern Patagonia Icefield (SPI). With an area of 13,000 km^2 and 48 major glaciers, the SPI is the largest ice mass in the Southern Hemisphere outside of Antarctica. The glacier inventory and recent glacier variations are presented, as well as ice thickness data and its variations, ice velocity, ablation, accumulation, hydrological characteristics, climate changes and implications for sea level rise. Most of the glaciers have been retreating, with a few in a state of equilibrium and advance. Glacier retreat is interpreted primarily as a response to regional atmospheric warming and to a lesser extent, to precipitation decrease observed during the last century in this region. The general retreat of SPI has resulted in an estimated contribution of 6% to the global rise in sea level due to melting of small glaciers and ice caps. Many glaciological characteristics of the SPI, in particular its mass balance, need to be determined more precisely.

2. INTRODUCTION

The Southern Patagonia Icefield (SPI) extends north-south for 370 km, between 48°15' S and 51°35' S, at an average longitude of 73°30' W (Figure 1). Its mean width is 35 km, and the minimum width is 9 km. The first detailed glacier inventory was compiled by Aniya et al. (1996), who showed that the SPI is composed of 48 major outlet glaciers and over 100 small cirque and valley glaciers. These glaciers flow from the Patagonian Andes to the east and west, generally terminating with calving fronts in freshwater lakes (east) and Pacific Ocean fjords (west).

[1] Centro de Estudios Científicos (CECS), Arturo Prat 514, Valdivia, Chile; [2] Universidad de Magallanes, Casilla 113-D, Punta Arenas, Chile; [3] Departamento de Geografía, Facultad de Arquitectura y Urbanismo, Universidad de Chile, Marcoleta 250, Santiago, Chile; [4] Institute of Geoscience, University of Tsukuba, Ibaraki 305-8571, Japan; [5] Institute of Low Temperature Science, Hokkaido University, Sapporo 060, Japan.

* corresponding author: gcasassa@cecs.cl

The Patagonian Icefields: A Unique Natural Laboratory for Environmental and Climate Change Studies.
Edited by Gino Casassa et al., Kluwer Academic /Plenum Publishers, 2002.

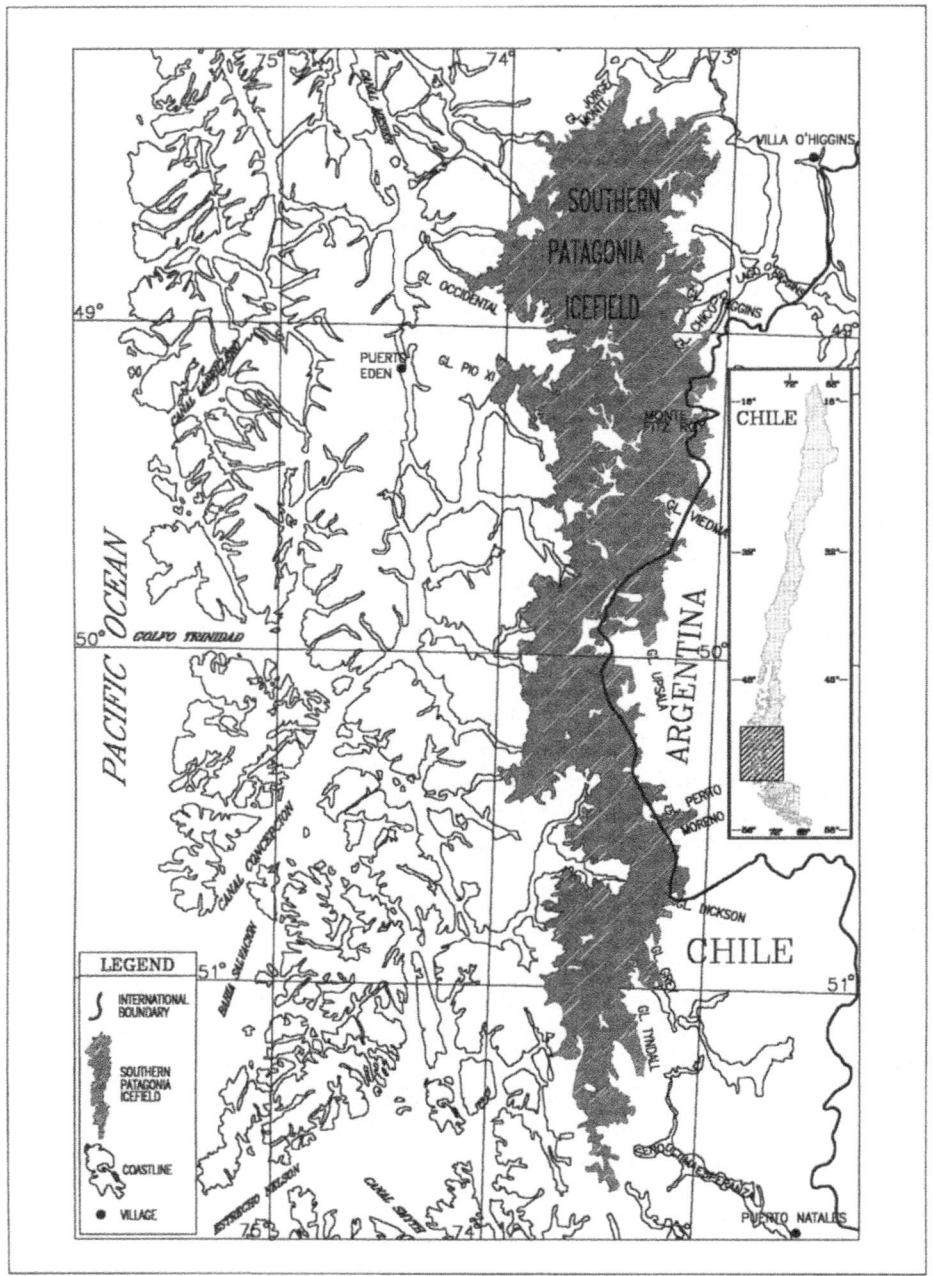

Figure 1. Location map, Southern Patagonia Icefield, Chile-Argentina. Modified from: Características Glaciológicas del Campo de Hielo Patagónico Sur, Casassa, G., Rivera, A., Aniya, M., and Naruse, R., 2000, *Anales del Instituto de la Patagonia, Serie Ciencias Naturales*, **28**:5-22. (a journal of the Universidad de Magallanes, Punta Arenas, Chile).

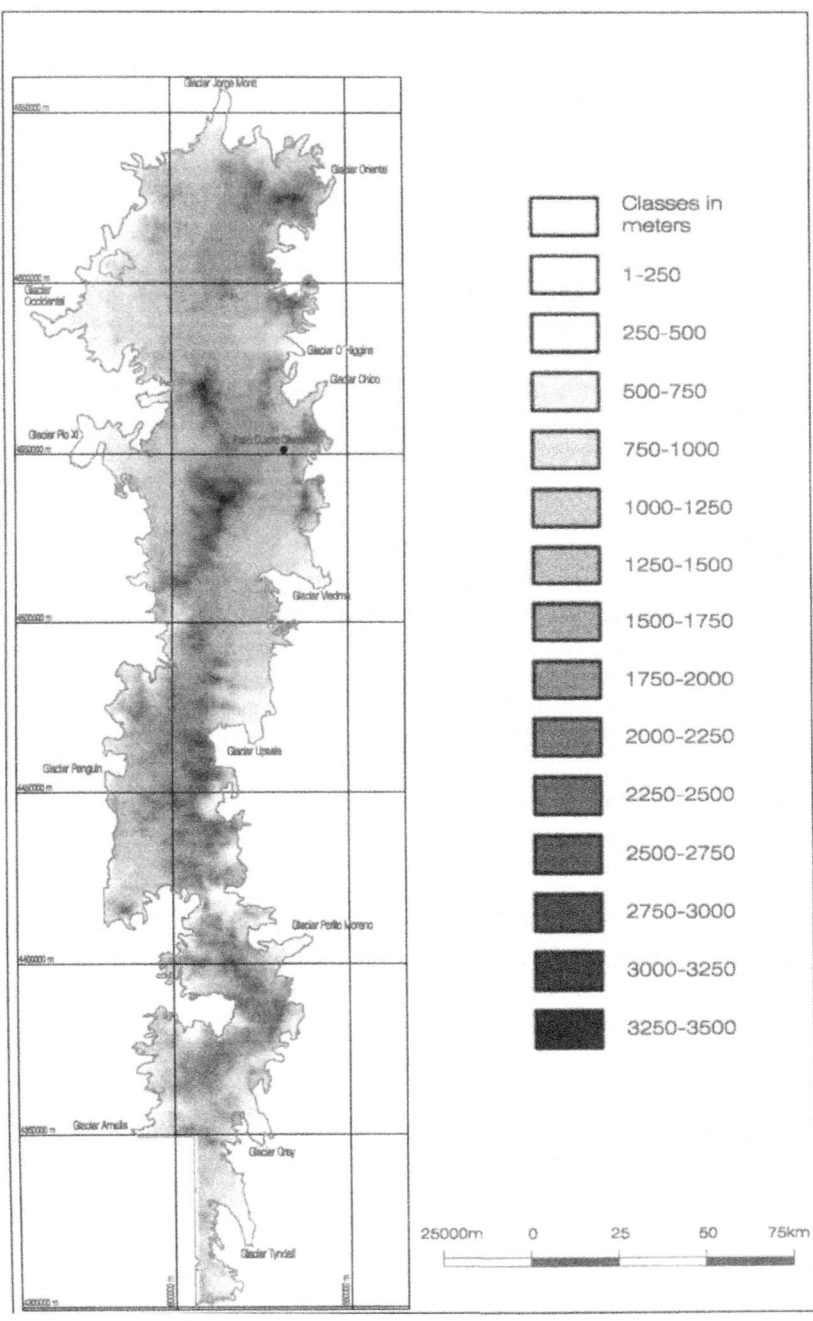

Figure 2. Digital terrain model for Southern Patagonia Icefield. Pixel size is 1 km. Data source are 1:250,000 preliminary maps, 1:100,000 maps from the Instituto Geográfico Militar, Argentina in the eastern SPI, and in the northern sector, 1:50,000 scale maps from the Instituto Geográfico Militar, Chile.

In the accumulation area, the outlet glaciers share a vast and relatively flat plateau with an average altitude of 1600 m (Figure 2). With a total area of 13,000 km^2 (Aniya *et al.*, 1996), the SPI is the largest ice mass in the Southern Hemisphere outside of Antarctica (Casassa *et al.*, 1998a).

According to the latest border agreement between Argentina and Chile, 81~92% of the SPI belongs to Chile and the rest belongs to Argentina. The official name for the SPI in Chile is "Campo de Hielo Sur" (Southern Icefield). The SPI is located 100 km south of the "Campo de Hielo Norte" (Northern Icefield), known in the scientific literature as the "Northern Patagonia Icefield" (NPI; Casassa, 1995). With an area of 4,200 km^2 (Aniya and Wakao, 1997; Aniya, 1988), the NPI is much smaller than the SPI, but their glaciological characteristics are very similar. In Argentina, the SPI is known as "Hielo Continental" (translated as "ice sheet"), a name that is not widely accepted because the SPI does not cover a large portion of a continent, but rather a reduced inland area. Other Spanish authors prefer "Hielo Patagónico Sur" (Southern Patagonia Ice: Lliboutry, 1956; Marangunic, 1964; Martinic, 1982), or "Campo de Hielo Patagónico Sur" (Southern Patagonia Icefield: Horvath, 1997; Casassa *et al.*, 2000). We prefer the latter name, because it preserves both a morphological and a geographical origin.

The SPI is located within the area affected by the southern westerlies. In consequence, due to a strong orographic effect, the western margin of the SPI receives a high amount of precipitation, with an average annual estimate of 10 m water equivalent (w.e.) for the icefield plateau (DGA, 1987; Casassa and Rivera, 1999). In contrast, the eastern margin of the SPI receives little precipitation, amounting to only a few hundred mm annually in the Argentine pampa (Ibarzabal y Donángelo *et al.*, 1996).

The mean annual temperature of the marginal areas of the SPI is approximately 6 °C (Carrasco *et al.*, 1998), which permits the existence of a unique ecosystem of beech forest at the glacier fronts. Due to the climatic regime, the ice in the SPI is temperate, at least in all of the ablation area and part of the accumulation area.

A comprehensive review of the glaciological studies performed at the Patagonian icefields was presented several years ago by Warren and Sugden (1993). Here we present an updated review of the SPI covering glacier inventory, glacier variations, ice thickness and its variations, ice velocity, mass balance, glacial hydrology, climate changes and sea level rise. A large part of this work has been previously published in Spanish (Casassa *et al.*, 2000).

3. GLACIER INVENTORY

Lliboutry (1956) was the first to describe the glaciological characteristics of the SPI. He studied Trimetrogon aerial photographs of 1944/45 and carried out field observations at de Los Tres Glacier, near Fitz Roy, determining a total area of 13,500 km^2 for the SPI. As part of the activities of the "Instituto Nacional del Hielo Continental Patagónico", Bertone (1960) compiled the first partial and preliminary inventory for the SPI, covering the glaciers that drain to lago Argentino. A few years later, the U.S. Army published a preliminary inventory for the whole SPI (Mercer, 1967).

Based on Landsat satellite imagery of 14 January 1986 (Figure 3), together with 1:250,000 "preliminary" topographic maps of the Instituto Geográfico Militar, Chile and stereoscopic analyses of aerial photographs, Aniya *et al.* (1996) compiled the complete inventory for the SPI (Table 1), identifying 48 major outlet glaciers.

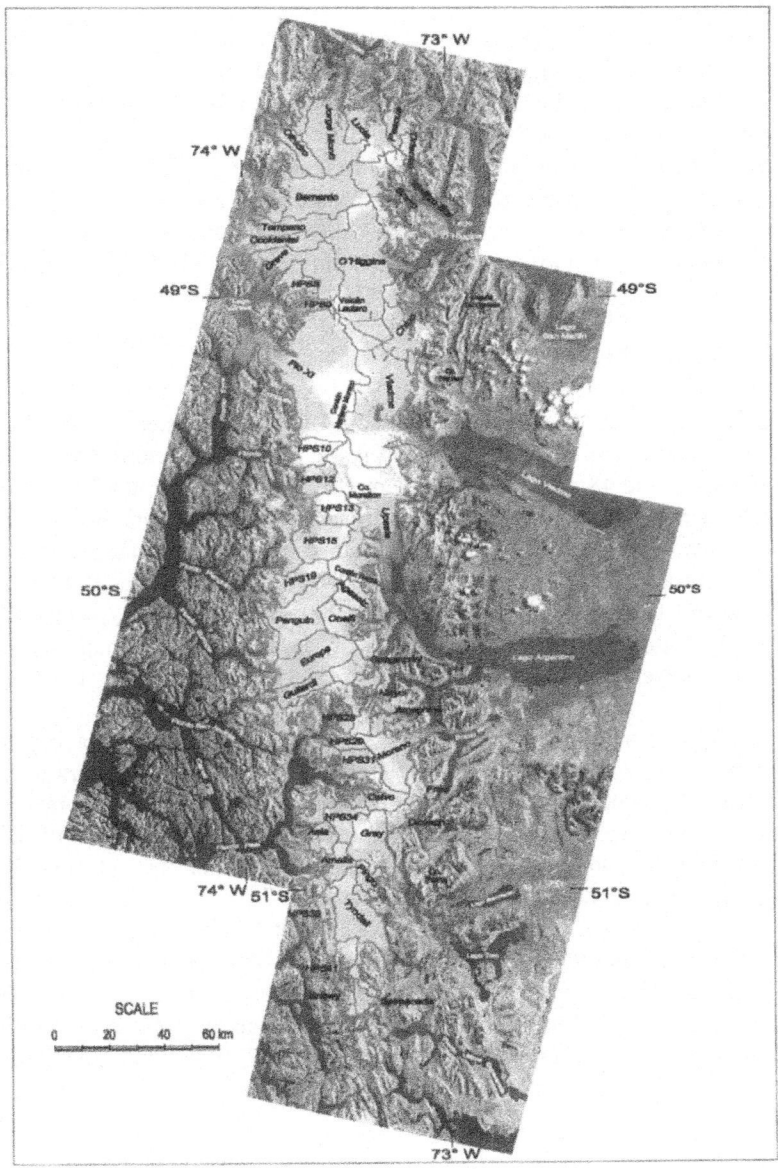

Figure 3. Satellite image mosaic of Southern Patagonia Icefield. The mosaic is a composite of three Landsat TM bands acquired 14 January 1986 (Naruse and Aniya, 1992). The mosaic was geolocated with 1:250,000 Preliminary Maps of Chile. Modified from: Características Glaciológicas del Campo de Hielo Patagónico Sur, Casassa, G., Rivera, A., Aniya, M., and Naruse, R., 2000, *Anales del Instituto de la Patagonia, Serie Ciencias Naturales*, **28**:5-22. (a journal of the Universidad de Magallanes, Punta Arenas, Chile).

Accumulation and ablation areas were estimated based on the snowline position on the Landsat imagery, assumed to coincide with the equilibrium line at the time of image acquisition. Of the 48 glaciers, 46 calve into freshwater lakes and fjords, while 2 terminate on land.

The largest glacier of the SPI is Pío XI, with a total area of 1265 km^2. Next largest glaciers are Viedma (945 km^2), Upsala (902 km^2) and O'Higgins (810 km^2). Areas are calculated by Aniya (personal communication), which are a better estimate of the values published by Aniya *et al.*, 1996.

The area covered by ice and snow within the 48 glaciers of the SPI is 11,259 km^2, with an additional 228 km^2 of exposed rocks in the accumulation area, which makes a total of 11,487 km^2. Adding to this value an area of 1,513 km^2 of small valley, cirque and mountain glaciers results in a total area of 13,000 km^2 for the SPI (Aniya *et al.*, 1996).

4. GLACIER VARIATIONS

Before 1997, information on glacier variations of the SPI existed for only half of the major glaciers (Warren and Sugden, 1993). Aniya *et al.* (1997), determined the area and frontal variations of the 48 major glaciers of the SPI, based on 1:250.000 "preliminary" maps of Chile, aerial photographs and satellite imagery. Other authors have extended the data set using historical data or recent aerial photographs of individual glaciers (Warren *et al.*, 1997; Rivera *et al.*, 2000; Rivera *et al.*, 1997a; Rivera *et al.*, 1997b; Agostini, 1945; Aniya *et al.*, 1999; Skvarca *et al.*, 1995a; 1999). In a few cases, glacier variations can be extended more than one century based on descriptions and maps produced by explorers (e.g. Naruse and Casassa, 1985; Casassa *et al.*, 1997a; Rivera, 1992; Martinic, 1999).

A generalized retreat is observed in 42 glaciers, while four glaciers were in equilibrium between 1944 and 1986 (HPS 13, HPS 15, Calvo and Spegazzini) and two advanced during the same period (Pío XI and Moreno). The largest retreat rate is shown by the O'Higgins Glacier, with a rate of loss in area of 1.21 km^2/a, or 484 m/a of frontal loss, between 1944/45 and 1985/86 (Aniya *et al.*, 1997). In the period 1896-1995, the O'Higgins Glacier retreated 14.6 km (Casassa *et al.*, 1997a).

The maximum advance was detected at the Pío XI Glacier (Mercer, 1964), with a rate of 1.45 km^2/a during the period 1946-1986, which translates into an average frontal advance of 288 m/a (Rivera *et al.*, 1997b). Recent field observations show that the Pío XI Glacier has been retreating since 1997. Perito Moreno Glacier is presently in equilibrium, but underwent frequent oscillations during the period 1947-1986, with a net gain in area of 4.1 km^2.

5. ICE THICKNESS

Casassa (1992) published ice thickness data for the SPI, detecting a maximum of 650 m of ice in the ablation area of Tyndall Glacier in 1990, using a 2.5 MHz analog radio-echo sounding device. In 1993, Casassa and Rivera (1998b) measured slightly smaller ice

Table 1. Glacier inventory for the Southern Patagonia Icefield. Glaciers are numbered counterclockwise from north to south. Adapted from Aniya *et al.* (1996). AAR represents accumulation area ratio divided by total glacier area. ELA is equilibrium line altitude.

Glacier	Latitude (S)	Longitude (W)	Length (km)	Total Area (km²)	Orient-ation	Accumul. Area (km²)	Ablation Area (km²)	AAR	ELA (m)	Calving Y/N	Max. Elevation (m)	Min. Elevation (m)
1 Jorge Montt	48° 04'	73° 30'	42	464	N	348	116	0.75	950	Y	2640	0
2 Ofhidro	48° 25'	73° 51'	26	116	NW	91	25	0.79	1000	Y	1655	45
3 Bernardo	48° 37'	73° 56'	51	536	W	444	92	0.83	1300	Y	2408	0
4 Témpano	48° 44'	74° 03'	47	332	W	242	90	0.73	900	Y	2408	0
5 Occidental	48° 51'	74° 14'	49	244	W	60	184	0.25	950	Y	---	<100
6 Greve	48° 58'	73° 55'	51	438	NW-W	292	146	0.67	1000	Y	3607	---
7 HPS8	49° 02'	73° 47'	11	38	SE	25	13	0.66	---	Y	---	---
8 HPS9	49° 03'	73° 48'	19	55	W	29	26	0.52	---	Y	3607	---
9 Pío XI	49° 13'	74° 00'	64	1265	W	1014	251	0.80	---	Y	3607	0
10 HPS10	49° 32'	73° 48'	16	61	W	---	---	---	---	Y	---	---
11 HPS12	49° 41'	73° 45'	23	204	S-W	164	40	0.80	---	Y	2257	0
12 HPS13	49° 43'	73° 40'	19	141	W	---	---	---	---	Y	2656	0
13 HPS15	49° 48'	73° 42'	19	174	N-W	164	10	0.94	---	Y	2446	0
14 HPS19	50° 00'	73° 55'	26	176	W	157	19	0.89	---	Y	---	0
15 Penguin	50° 05'	73° 55'	38	527	NW	507	20	0.96	---	Y	3180	0
16 Europa	50° 18'	73° 52'	39	403	W	379	24	0.94	---	Y	---	0
17 Guilardi	50° 23'	73° 57'	36	148	W	125	23	0.85	---	Y	---	0
18 HPS28	50° 25'	73° 35'	12	63	W	47	16	0.75	---	Y	2238	0
19 HPS29	50° 28'	73° 36'	17	82	W	69	13	0.85	1200	Y	2950	0
20 HPS31	50° 36'	73° 33'	23	161	SW	141	20	0.88	900	Y	2950	0
21 Calvo	50° 41'	73° 21'	13	117	W	114	3	0.97	---	Y	---	0
22 HPS34	50° 43'	73° 32'	14	137	NW	122	15	0.89	800	Y	---	0
23 Asia	50° 49'	73° 44'	12	133	W	86	47	0.65	---	Y	2179	0
24 Amalia	50° 57'	73° 45'	21	158	W	126	32	0.80	900	Y	---	0
25 HPS38	51° 03'	73° 45'	16	62	W	27	35	0.44	---	Y	---	---
26 HPS41	51° 18'	73° 34'	17	71	SW	39	32	0.55	---	Y	---	---

Table 1. continues on next page

Table 1 (continued).

Glacier	Latitude (S)	Longitude (W)	Length (km)	Total Area (km²)	Orient- ation	Accummul. Area (km²)	Ablation Area (km²)	AAR	ELA (m)	Calving Y/N	Max. Elevation (m)	Min. Elevation (m)
27 Snowy	51° 22'	73° 34'	9	23	W	11	12	0.48	---	Y	---	---
28 Balmaceda	51° 23'	73° 18'	12	63	E	42	21	0.67	650	Y	---	---
29 Tyndall	51° 15'	73° 15'	32	331	E	213	118	0.64	900	Y	---	50
30 Pingo	51° 02'	73° 21'	11	71	SE	56	15	0.79	---	Y	---	200
31 Grey	51° 01'	73° 12'	28	270	SE	167	103	0.62	---	Y	---	100
32 Dickson	50° 47'	73° 09'	10	71	SE	42	29	0.59	---	Y	---	---
33 Frías	50° 45'	75° 05'	9	48	E	30	18	0.62	---	N	---	---
34 Moreno	50° 30'	73° 00'	30	258	NE	188	70	0.73	1150	Y	2950	175
35 Ameghino	50° 25'	73° 10'	21	76	N	32	44	0.42	1000	Y	2250	201
36 Mayo	50° 22'	73° 20'	15	45	N-S	28	17	0.62	900	Y	2250	200
37 Spegazzini	50° 15'	73° 20'	17	137	E-S	116	21	0.85	---	Y	2940	175
38 Onelli	50° 07'	73° 25'	13	84	NE-S	52	32	0.62	---	Y	3064	175
39 Agassiz	50° 06'	73° 22'	17	50	E	37	13	0.74	---	Y	3180	175
40 Upsala	49° 59'	73° 17'	60	902	SE	611	290	0.68	1150	Y	---	250
41 Viedma	49° 31'	73° 01'	71	945	E-S	564	381	0.60	1250	Y	---	285
42 Chico	49° 00'	73° 04'	25	243	E	194	49	0.80	---	Y	---	285
43 O'Higgins	48° 55'	73° 08'	46	810	N-E-S	701	109	0.87	1300	Y	3607	285
44 Bravo	48° 38'	73° 10'	23	129	E	98	31	0.76	1500	N	3067	300
45 Mellizo Sur	48° 37'	73° 07'	14	37	SE	32	5	0.86	1400	Y	3067	300
46 Oriental	48° 27'	73° 01'	17	74	E	56	18	0.75	1150	Y	3017	285
47 Pascua	48° 22'	73° 09'	23	88	N	58	30	0.66	950	Y	3017	151
48 Lucía	48° 20'	73° 20'	29	200	N	145	55	0.72	1000	Y	3067	27
TOTAL			11,259			8,285	2,773	0.75				

Notes: Glacier 23 (Asia according to Lliboutry, 1956) is named "Brujo" in the 1:100,000 map "Península Wilcock" of IGM Chile.
Glacier 24 (Bravo according to the 1:50,000 map "Mellizo Sur" of IGM Chile), is locally known as "Rivera". Lliboutry (1956) also names it "Rivera".
Moreno Glacier is also known as "Perito Moreno" Glacier.

thickness values in the same area of Tyndall Glacier, this time with a digital radio-echo sounding system, concluding that the glacier had thinned in the period 1990-1993.

In 1995 and 1996, Casassa *et al.* (1997b) performed ice thickness measurements with a digital radar system near the front of Grey Glacier. In 1997, Rivera and Casassa (2000) performed ice thickness measurements at Paso de los Cuatro Glaciares by means of a profiling radar system pulled by a snowmobile, measuring ice thicknesses in excess of 750 m, the maximum range of the radar.

Rott *et al.* (1998) measured ice thickness on Moreno Glacier using seismic reflection and explosive charges along a transect on the ablation area. In addition to ice thickness, Rott and co-workers could also detect layers of subglacial sediments. In 1999 and 2000, researchers from the University of Washington, U.S.A., performed ice thickness measurements in the ablation area of Tyndall Glacier by means of a digital radar system, detecting a maximum of 650 m (Raymond *et al.*, 2000). Rivera and Casassa (2002; Chapter 10) present a summary of ice thickness measurements at SPI.

6. ICE THICKNESS CHANGES

Several measurements of ice thickness changes have been carried out at the ablation areas based on repeated ground surveys and a comparison of glacier maps corresponding to different dates. The results indicate that a regional thinning prevails, with one glacier showing no variation (Moreno), and one glacier which thickened (Pío XI). The results are shown in Table 2.

Table 2. Ice thickness changes for SPI.

GLACIER NAME	RATE (m a^{-1}) THINNING (-) THICKENING (+)	PERIOD	REFERENCE
O'Higgins	-3.2	1914 - 1933	Casassa *et al.*, 1997a
	-6.7	1933 - 1960	Casassa *et al.*, 1997a
	-2.5 to -11	1975 - 1995	Casassa *et al.*, 1997a
Pío XI	+2.2	1975 - 1995	Rivera and Casassa, 1999
Upsala	-3.6	1968 - 1990	Aniya *et al.*, 1997
	-9.5 to -14	1991 - 1993	Naruse *et al.*, 1995a, Skvarca *et al.*, 1995b, Naruse *et al.*, 1997
Ameghino	-2.3	1949 - 1993	Aniya 1999
Moreno	No change	1991 - 1993	Skvarca and Naruse, 1997 Naruse *et al.*, 1995b
Dickson	-2.5 to -8.1	1975 - 1998	Rivera *et al.*, 2000
Grey	-2.3	1975 - 1995	Casassa *et al.*, 1997b
Tyndall	-2.0	1945 - 1993	Aniya *et al.*, 1997
	-1.7	1975 - 1985	Kadota *et al.*, 1992
	-4.0	1985 - 1990	Kadota *et al.*, 1992
	-3.1	1990 - 1993	Nishida *et al.*, 1995

Ranges of thickness changes correspond to spatially different observations within the ablation area.

7. ICE VELOCITY

Most of the ice velocity measurements in the SPI have been carried out in the ablation area of the glaciers with traditional surveying methods.

Satellite techniques have been applied more recently for determining ice velocity, including both radar interferometry and phase correlation of radar data (Rott *et al.*, 1998; Michel and Rignot 1999; Forster *et al.*, 1999).

The largest velocity values have been measured in late spring at the calving front of Pío XI Glacier, with a maximum value of 50 m/d and a mean value of 20 m/d for a 3-day period (Rivera *et al.*, 1997b). Velocity data are presented in Table 3.

Table 3. Ice velocity measurements at SPI.

GLACIER NAME	VELOCITY m/d	MEASUREMENT PERIOD	MEASUREMENT METHOD	REFERENCE
Upsala	4.44	21-29 November, 1993	Theodolite	Skvarca *et al.*, 1995b
	3.7	14-18 November, 1990	Theodolite	Naruse *et al.*, 1992
Moreno	1.1 to 2.19	9-10 October, 1994	Interferometry	Michel and Rignot, 1999
	2.1 to 5	14-18 November, 1990	Theodolite	Naruse *et al.*, 1992
	0.5 to 3.5	October, 1994	Interferometry	Rott *et al.*, 1998
	2.64	November, 1993 to December, 1994	Theodolite	Skvarca and Naruse, 1997
Tyndall	0.1 to 1.9	30 November, 1985 to 3 December, 1985	Theodolite	Naruse *et al.*, 1987
	0.07 to 0.51	7-15 December, 1990	Theodolite	Kadota *et al.*, 1992
	0.065 to 0.61	9-18 December, 1993	Theodolite	Nishida *et al.*, 1995
Penguin	0.9 to 2.2	October, 1994	Interferometry	Forster *et al.*, 1999
Pío XI	1 to 50	14-17 November, 1995	Theodolite	Rivera *et al.*, 1997b

Ranges of velocities correspond to observations at different sites on each glacier.

8. MASS BALANCE

Very limited data of ablation and accumulation exists for the SPI. Based on meteorological and hydrological data available for stations located around the SPI, isolines of precipitation have been estimated (DGA, 1987), and general characteristics of glacial hydrology of SPI have been published (Peña and Escobar, 1987; Peña and Gutiérrez, 1992).

Space imagery can provide information about snow and ice surface patterns (Forster *et al.*, 1996), but at SPI it has not yet been possible to quantify ablation and accumulation based on such imagery. Physical models based exclusively on climatic conditions can provide an estimate of mass balance, both during present and past conditions (Hulton and Sugden, 1995), but accuracy of such results can be poor if no calibration with field data is available.

Ablation data have been obtained in the following glaciers by the traditional stake method for periods ranging from a few days to a few months:

- Tyndall Glacier: Takeuchi *et al.*, 1995, Koizumi and Naruse, 1992.
- Moreno Glacier: Skvarca and Naruse, 1997.
- Pío XI Glacier: Casassa and Rivera, 1999.

The only annual ablation-stake measurement has been done at Moreno Glacier, with a magnitude of 10 m/a (Skvarca and Naruse, 1997), a value considered to be representative of the lower ablation area of the SPI. A similar value was calculated based on ablation data for a period of a few months and extrapolated using the degree-day method (Takeuchi *et al.*, 1995).

As for accumulation, only one field measurement using the stake method is available for annual periods, corresponding to a site at 1460 m elevation on Chico Glacier where a mean accumulation of 0.78 m/a water equivalent (w.e.) was measured with a 12 m-high tower in the period 1996-1998 in combination with snow pit data (Casassa and Rivera, 1999). A few accumulation data based on snow pit studies and short-period snow stake heights are available for specific sites on the SPI (Casassa and Rivera, 1999).

Two shallow firn cores have been retrieved at the SPI. The first one corresponds to the ice divide at Moreno Glacier, with a length of 13.2 m (Aristarain and Delmas, 1993), who estimated an annual accumulation of 1.2 m/a w.e. The second core corresponds to the ice divide at Tyndall Glacier, with a length of 45.97 m and an annual precipitation estimate of 13,5 m/a w.e. (Godoi *et al.*, 2002; Chapter 14).

Based on the scarce ablation and accumulation data available, Casassa and Rivera (1999) developed a digital topographic model at a resolution of 1 km for 98% of the total area of SPI, leaving out of the study area the south western portion of SPI, where no topographic information was available. Their results show an annual ablation of 23.8 km^3/a over an ablation area of 4100 km^2 and an annual precipitation of 58.4 km^3/a over an accumulation area of 8700 km^2. A residual value of 34.6 km^3/a results if steady state is assumed. This represents the volume wasted from 98% of the area of SPI, which should calve as icebergs into lakes and fjords.

Two studies have used the hydrometeorological method to compute mass balances for the SPI. In the basin of Serrano river, Marangunic (personal communication) calculated a mass balance by applying algorithms for estimating ablation and accumulation rates at different elevations. Escobar *et al.* (1992) calculated a hydrological balance for the SPI based on the climatic and hydrologic data published by DGA (1987).

9. GLACIAL FLOODS

Several cases of glacial floods, known as *jökulhlaup*, have occurred at the SPI. At Dickson Glacier, a flood associated with the sudden drainage of a proglacial lake was observed in 1982 and 1987. The flood of 1982 at Dickson Glacier was described and modeled by Peña and Escobar (1985). Both floods affected the hydrological system downstream from Dickson Glacier, increasing river and lake levels by several meters. At present, due to the large retreat of Dickson Glacier, the proglacial lake is now permanently joined to Dickson lake, so that a new *jökulhlaup* is not expected in the near future.

The most spectacular case of *jökulhlaup* in the SPI is that of Moreno Glacier, where several flood events have occurred during the last century (Nichols and Miller, 1952; Mercer, 1968; Liss, 1970). The floods are produced by the periodic advance of Moreno Glacier over the Magallanes península, effectively damming the Brazo Sur of Lago Argentino, known as Brazo Rico. A maximum lake level rise exceeding 10 m has been detected, evidenced by old coast lines preserved around Brazo Rico. When the lake level of Brazo Rico compared to the main water body of Lago Argentino reaches a certain critical level, Brazo Rico drains catastrophically, firstly through an englacial tunnel and later by an open-water channel. The last *jökulhlaup* occurred in the mid-1980's. Since then, Moreno Glacier has not dammed Brazo Rico again, and a narrow open water canal exists at the glacier front, continuously draining Brazo Rico.

10. CLIMATE CHANGES IN CHILE

Historical records of air temperature in Chile have been analyzed by Rosenblüth *et al.* (1995) and Rosenblüth *et al.* (1997), detecting a warming rate of 1.3 to 2.0 °C/100 years in the period 1933-1992. This warming has practically doubled at some stations in western Patagonia during the last three decades (Rosenblüth *et al,* 1997). Climate studies of the eastern margin of the SPI indicate that this warming also prevails in Argentine Patagonia (Ibarzabal y Donángelo *et al.,* 1996).

A few Chilean and Argentine stations around the SPI show an important precipitation decrease of approximately 25-33% during the last century (Rosenblüth *et al.,* 1997, Ibarzabal y Donángelo *et al.,* 1996), although other stations show no change (Santana, 1984). Recent data show a precipitation increase at some stations (Carrasco *et al.,* 2002; Chapter 4).

The climatic data suggest that the generalized glacier retreat in the SPI, is due to a regional warming, together with a probable precipitation decrease during the last century in the area.

11. SEA LEVEL RISE

Sea level has been rising in the last 100 years at a rate of 1 to 2 mm/a (Gornitz, 1995). This sea level rise is related to several causes, including thermal expansion due to global warming of the oceans, melting of glaciers and small ice caps, and the contribution of the Antarctic and Greenland ice sheets. Warrick and Oerlemans (1990) estimated that glaciers and small ice caps presently contribute 0.4 mm/a of the total sea level rise.

A few authors have related the glacier variations of the Patagonian glaciers to global sea level change. Meier (1984) calculated that the world-wide contribution of small glaciers and ice caps (excluding Greenland and Antarctica) to sea level between 1900 and 1961, was 0.46 ± 0.26 mm/a. Meier further estimated that the Andes south of 30° S accounted as a whole, for about 12% of this value, that is, 0.06 mm/a, without calculating a separate contribution for the Patagonian icefields due to lack of data. Dyurgerov and

Meier (1997) estimated a world-wide sea-level rise of 0.25 ± 0.10 mm/a due to small glaciers, but excluded data of the Patagonian icefields, so that the calculation of Meier (1984) is considered to be more representative.

Aniya (1999) calculated an ice volume loss of 594 ± 239 km^3 for the SPI in the period 1945-1996, obtaining a water equivalent volume of 505 ± 203 km^3. Rivera *et al.* (in press) calculated a total water equivalent volume loss of 401 ± 174 km^3 for the SPI in 1945-1996. Rivera's estimate is smaller than Aniya's because a smaller glacier area for SPI was used, limited to the area of the outlet glaciers, and a reduced area and volume loss measured at Pio XI Glacier was considered. Dividing the water equivalent value of Rivera and co-workers by the total ocean area of the world, results in a contribution of 1.108 ± 0.481 mm/a in the 51 year-period from 1945 to 1996, that is, 0.022 ± 0.009 mm/a. If this is valid, then the SPI contributes about 6% of the total sea level rise due to small glaciers and ice caps as estimated by Meier (1984), or 10% if all the Chilean glaciers are considered (Rivera *et al.*, in press).

12. CONCLUSIONS

Field data, complemented by information recovered from aerial photography and satellite imagery, have provided a glacier inventory for the SPI and the variations of its major glaciers since 1945. Except for a few cases, the glaciers show a general retreat. The variation record has been extended back to the XIXth century in the case of a few glaciers such as Pío XI and O'Higgins. The retreat is probably caused by a regional warming observed during the last 100 years, together with an apparent precipitation decrease. The advance of Pío XI and Moreno Glaciers could be due to dynamic or topographic causes of a local nature, or simply a non-linear response of calving glaciers to climate (Warren and Rivera, 1994). The general retreat observed at the SPI results in a total estimated contribution of 6% to the global sea level rise due to melting of small glaciers and ice caps.

The use of traditional and modern field techniques together with remote sensing methods have added important glaciological information, such as ice velocity, ice thickness, and ablation/accumulation, for some areas of the SPI. However, lack of spatial distribution of such glaciological information, together with the limited availability of regular maps at a scale of 1:100.000 and 1:50.000, prevents a more detailed characterization of the SPI, particularly of its accumulation area. Specific mass balance and ice flux data are needed, which would permit a more precise contribution to sea-level rise and adequate calibration for modeling the present and future behavior of SPI.

13. ACKNOWLEDGMENTS

Part of this work was funded by Fondecyt project 1980293 and by Grants in Aid for International Scientific Research Program by the Japanese Ministry of Education, Science and Culture (1990: No. 02041004 and 1993: No. 05041049). We thank César Acuña for drafting the maps and María Angélica Godoi, for help with Table 1. We also appreciate valuable comments from Mr. Mateo Martinic and Al Rasmussen. Centro de

Estudios Científicos is a Millennium Science Institute. This article is a modified version of an article originally published in Spanish: Características glaciológicas del Campo de Hielo Patagónico Sur, Casassa *et al.*, 2000, *Anales del Instituto de la Patagonia,* Serie Ciencias Naturales, **28**:5-22 (a journal of the Universidad de Magallanes, Punta Arenas, Chile).

14. REFERENCES

Agostini, A., 1945, *Andes Patagónicos. Viajes de Exploración a la Cordillera Patagónica Austral,* 2nd ed., Guillermo Kraft, Buenos Aires, 445 pp. (in Spanish).

Aniya, M., 1988, Glacier inventory for the Northern Patagonia Icefield, Chile, and variations 1944/45 to 1985/86, *Arctic and Alpine Research,* **20**:179-187.

Aniya, M., 1999, Recent glacier variations of the Hielos Patagónicos, South America, and their contribution to sea-level change, *Arctic and Alpine Research,* **31**(2):165-173.

Aniya, M., Sato, H., Naruse, R., Skvarca, P., and Casassa, G., 1996, The use of satellite and airborne imagery to inventory outlet glaciers of the Southern Patagonia Icefield, South America, *Photogrammetric Engineering and Remote Sensing,* **62**:1361-1369.

Aniya, M., and Wakao., Y., 1997, Glacier variations of Hielo Patagónico Norte, Chile, between 1944/45 and 1995/96, *Bulletin of Glacier Research,* **15**:11-18.

Aniya, M., Sato, H., Naruse, R., Skvarca, P., and Casassa, G., 1997, Recent variations in the Southern Patagonia Icefield, South America, *Arctic and Alpine Research,* **29**(1):1-12.

Aniya, M., Naruse, R., Casassa, G., and Rivera, A., 1999, Variations of Patagonian glaciers, South America, utilizing RADARSAT images, Proceedings of the International Symposium on RADARSAT Application Development and Research Opportunity (ADRO), Montreal, Canada, October 13-15, 1998, CD-ROM.

Aristarain, A., and Delmas, R., 1993, Firn-core study from the Southern Patagonia ice cap, South America, *Journal of Glaciology,* **39**(132):249-254.

Bertone, M., 1960, Inventario de los glaciares existentes en la vertiente Argentina entre los paralelos 47°30' y 51° S, Publication. N° 3, Instituto Nacional del Hielo Continental Patagónico, Buenos Aires, 103 pp. (in Spanish).

Carrasco, J., Casassa, G., and Rivera, A., 1998, Climatología actual del Campo de Hielo Sur y posibles cambios por el incremento del efecto invernadero, *Anales Instituto de la Patagonia, Serie Ciencias Naturales,* **26**:119-128.

Carrasco, J. F., Casassa, G., and Rivera, A., 2002, Meteorological and climatogical aspects of the Southern Patagonia Icefield, in: *The Patagonian Icefields: a unique natural laboratory for environmental and climate change studies,* G. Casassa, F. V. Sepúlveda, and R. M. Sinclair, eds., Kluwer Academic/Plenum Publishers, New York, pp. 29-41.

Casassa, G., 1992, Radio-echo sounding of Tyndall Glacier, Southern Patagonia, *Bulletin of Glacier Research,* **10**: 69-74.

Casassa, G., 1995, Glacier inventory in Chile: current status and recent glacier variations, *Annals of Glaciology,* **21**:317-322.

Casassa, G., Brecher, H., Rivera, A., and Aniya, M., 1997a, A century-long record of glacier O'Higgins, Patagonia, *Annals of Glaciology,* **24**:106-110.

Casassa, G., Rivera, A., Lange, H., and Carvallo, R., 1997b, Retreat of Grey glacier: a response to regional warming in Patagonia, Abstracts 1997 Joint Assemblies of the International Association of Meteorology and Atmospheric Sciences (IAMAS) and the International Association for Physical Sciences of the Oceans (IAPSO), Melbourne, 1-9 July, 1997, JMPH18-11.

Casassa, G., Espizúa, L., Francou, B., Ribstein, P., Ames, A., and Alean, J., 1998a, Glaciers in South America, in: *Into the Second Century of World Wide Glacier Monitoring: Prospects and Strategies,* Haeberli, Hoelzle and Suter, eds., World Glacier Monitoring Service, UNESCO *Studies and Reports in Hydrology,* **56**:125-146, Zürich.

Casassa, G., and Rivera, A., 1998b, Digital radio-echo sounding at Tyndall glacier, Patagonia, *Anales Instituto de la Patagonia, Serie Ciencias Naturales,* **26**:129-135.

Casassa, G., and Rivera, A., 1999, Topographic mass balance model for the Southern Patagonia Icefield, Abstracts International Symposium on the Verification of Cryospheric models, Bringing data and modelling scientists together, 16-20 August 1999, Zürich, p. 44.

Casassa, G., Rivera, A., Aniya, M., and Naruse, R., 2000, Características glaciológicas del Campo de Hielo Patagónico Sur, *Anales Instituto de la Patagonia, Serie Ciencias Naturales (Chile)*, **28**:5-22. (in Spanish).

Dirección General de Aguas (DGA), 1987, *Balance Hídrico de Chile*, Ministerio de Obras Públicas, Santiago, Chile, 59 pp. (in Spanish).

Dyurgerov, M. B., and Meier, M. F., 1997, Year-to-year fluctuation of global mass balance of small glaciers and their contribution to sea-level changes, *Arctic and Alpine Research*, **29**:392-402.

Escobar, F., Vidal, F., and Garín, C., 1992, Water balance in the Patagonia Icefield, in: *Glaciological Researches in Patagonia, 1990*, R. Naruse and M. Aniya, eds., Japanese Society of Snow and Ice, pp. 109-119.

Forster, R., Isacks, B., and Das, D., 1996, Shuttle imaging radar (SIR-C/X-SAR) reveals near-surface properties of the south Patagonian ice-field, *Journal of Geophysical Research*, **101**(E10):23169-23180.

Forster, R., Rignot, E., Isacks, B., and Jezek, K., 1999, Interferometric radar observations of glaciares Europa and Penguin, Hielo Patagónico Sur, Chile, *Journal of Glaciology*, **45**(150):325-337.

Godoi, M. A., Shiraiwa, T., Kohshima, S., and Kubota, K., 2002, Firn-core drilling operation at Tyndall glacier, Southern Patagonia Icefield, in: *The Patagonian Icefields: a unique natural laboratory for environmental and climate change studies*, G. Casassa, F. V. Sepúlveda, and R. M. Sinclair, eds., Kluwer Academic/Plenum Publishers, New York, pp. 149-156.

Gornitz, V., 1995, Sea-level rise: a review of recent past and near-future trends, *Earth Surface Processes and Landforms*, **20**:7-20.

Horvath, A., 1997, *La Definición de Límites o el Límite a la Indolencia*. Ediciones Cruz del Sur de la Trapananda, Coihaique, pp. 131 (in Spanish).

Hulton, R., and Sugden, D., 1995, Modelling mass balance on former maritime ice caps: a Patagonian example, *Annals of Glaciology*, **21**:304-310.

Ibarzabal y Donángelo, T., Hoffmann, J., and Naruse, R., 1996, Recent climate changes in southern Patagonia, *Bulletin of Glacier Research*, **14**:29-36.

Kadota, T., Naruse, R., Skvarca, P., and Aniya, M., 1992, Ice flow and surface lowering of Tyndall Glacier, Southern Patagonia, *Bulletin of Glacier Research*, **10**:63-68.

Koizumi, K., and Naruse, R., 1992, Measurements of meteorological conditions and ablation at Tyndall Glacier, Southern Patagonia, in December 1990, *Bulletin of Glacier Research* (Japanese Society of Snow and Ice), **10**:79-82.

Liss, C. C., 1970, Der Moreno Gletscher in der Patagonischen Kordillere, *Zeitschrift für Gletscherkunde und Glazialgeologie*, Bd. **6**, H.1-2:161-180 (in German).

Lliboutry, L., 1956, *Nieves y Glaciares de Chile. Fundamentos de Glaciología*, Ediciones de la Universidad de Chile, Santiago, pp. 471 (in Spanish).

Marangunic, C., 1964, *Observaciones Glaciológicas y Geológicas en la Zona del Paso de los Cuatro Glaciares, Hielo Patagónico Sur*, Undergraduate Thesis (Geology), Universidad de Chile, Santiago, pp. 125 (in Spanish).

Martinic, M., 1982, *Hielo Patagónico Sur*, Ediciones Instituto de la Patagonia, Punta Arenas, pp. 119 (in Spanish).

Martinic, M., 1999, *Cartografía Magallánica 1523-1945*, Ediciones de la Universidad de Magallanes, Punta Arenas, pp. 345 (in Spanish).

Meier, M., 1984, Contribution of small glaciers to global sea level, *Science*, **226**:1418-1420.

Mercer, J. H., 1964, Advance of a Patagonian glacier, *Journal of Glaciology*, **5**:267-268.

Mercer, J. H., 1967, *Southern Hemisphere Glacier Atlas*, U.S. Army, Natick Laboratories, Technical report 67-76-ES, Massachussetts, pp. 325.

Mercer, J. H., 1968, Variations of some Patagonian glaciers since the Late-Glacial, *American Journal of Science*, **266**:91-109.

Michel, R., and Rignot, E., 1999, Flow of glaciar Moreno, Argentina, from repeat-pass Shuttle Imaging Radar images: comparison of the phase correlation method with radar interferometry, *Journal of Glaciology*, **45**(149):93-100.

Naruse, R., and Casassa, G., 1985, Reconnaissance survey of some glaciers in the Southern Patagonia Icefield, in: *Glaciological Studies in Patagonia Northern Icefield, 1983-84*, C. Nakajima, ed., Data Center for Glacier Research, Japanese Society of Snow and Ice, pp. 121-133.

Naruse, R., and Aniya, M., 1992, Outline of Glacier Research Project in Patagonia, 1990, *Bulletin of Glacier Research*, **10**:31-38.

Naruse, R., Peña, H., Aniya, M., and Inoue, J., 1987, Flow and surface structure of Tyndall glacier, Southern Patagonia Icefield, *Bulletin of Glacier Research*, **4**:133-140.

Naruse, R., Skvarca, P., Kadota, T., and Koizumi, K., 1992, Flow of Upsala and Moreno glaciers, Southern Patagonia, *Bulletin of Glacier Research*, **10**:55-62.

Naruse, R., Aniya, M., Skvarca, P., and Casassa, G., 1995a, Recent variations of calving glaciers in Patagonia, South America, revealed by ground surveys, satellite-data analyses and numerical experiments, *Annals of Glaciology,* **21**:297-303.

Naruse, R., Skvarca, P., Satow, K., Takeuchi, Y., and Nishida, K., 1995b, Thickness change and short-term flow variation of Moreno glacier, Patagonia, *Bulletin of Glacier Research,* **13**:21-28.

Naruse, R., Skvarca, P., and Takeuchi, Y., 1997, Thinning and retreat of Upsala Glacier, and an estimate of annual ablation changes in Southern Patagonia, *Annals of Glaciology,* **24**:38-42.

Nichols, N. L., and Miller, M. M., 1952, Advancing glaciers and nearby simultaneously retreating glaciers, *Journal of Glaciology,* **2**(11):41-50.

Nishida, K., Satow, K., Aniya, M., Casassa, G., and Kadota, T., 1995, Thickness change and flow of Tyndall Glacier, Patagonia, *Bulletin of Glacier Research,* **13**:29-34.

Peña, H., and Escobar, F., 1985, *Análisis de las crecidas del río Paine, XII Región.* Publicación Interna Estudios Hidrológicos Numero **83/7**, Dirección General de Aguas, Departamento de Hidrología, Santiago, pp. 78. (in Spanish).

Peña, H., and Escobar, F., 1987, Aspects of glacial hydrology in Patagonia, *Bulletin of Glacier Research,* **4**:141-150.

Peña, H., and Gutiérrez, R., 1992, Statistical analysis of precipitation and air temperature in the Southern Patagonia Icefield, in: *Glaciological Researches in Patagonia, 1990,* R. Naruse and M. Aniya, eds., Japanese Society of Snow and Ice, pp. 95-107.

Raymond, C., Neuman, T., Rignot, E., Rivera, A., and Casassa, G., 2000, Retreat of Tyndall glacier, Patagonia, Chile, in: *EOS, Transactions, American Geophysical Union,* **81**(48):F427, H61G-02.

Rivera, A., 1992, El glaciar Pío XI: avances y retrocesos, el impacto sobre su entorno durante el presente siglo, *Revista Geográfica de Chile Terra Australis,* **36**:33 - 62. (in Spanish).

Rivera, A., Aravena, J., and Casassa, G., 1997a, Recent fluctuations of glaciar Pío XI, Patagonia: discussion of a glacial surge hypothesis, *Mountain Research and Development,* **17**(4):309-322.

Rivera, A., Lange, H., Aravena, J., and Casassa, G., 1997b, The 20th century advance of glaciar Pío XI, Southern Patagonia Icefield, *Annals of Glaciology,* **24**: 66-71.

Rivera, A., and Casassa, G., 1999, Volume changes of Pío XI glacier:1975-1995, *Global Planetary Change,* **22**(1-4):233-244.

Rivera, A., and Casassa, G., 2000, Variaciones recientes del glaciar Chico, Campo de Hielo Sur, Actas IX Congreso Geológico Chileno, Puerto Varas, 31 Julio-4 Agosto 2000, pp. 244-248 (in Spanish).

Rivera, A., and Casassa, G., 2002, Ice thickness measurements on the Southern Patagonia Icefield, in: *The Patagonian Icefields: a unique natural laboratory for environmental and climate change studies,* G. Casassa, F. V. Sepúlveda, and R. M. Sinclair, eds., Kluwer Academic/Plenum Publishers, New York, pp. 101-115.

Rivera, A., Casassa, G., Acuña, C., and Lange, H., 2000, Variaciones recientes de glaciares en Chile, *Revista Investigaciones Geográficas,* **34**:29-60 (in Spanish).

Rivera, A., Acuña, C., Casassa, G., and Brown, F., in press, Use of remotely sensing and field data to estimate the contribution of Chilean glaciers to eustatic sea level rise, *Annals of Glaciology,* **34**.

Rosenblüth, B., Casassa, G., and Fuenzalida, H., 1995, Recent climate changes in western Patagonia, *Bulletin of Glacier Research,* **13**:127-132.

Rosenblüth, B., Fuenzalida, H., and Aceituno, P., 1997, Recent temperature variations in southern South America, *International Journal of Climatology,* **17**:67-85.

Rott, H., Stuefer, M., Siegel, A., Skvarca, P., and Eckstaller, A., 1998, Mass fluxes and dynamics of Moreno Glacier, Southern Patagonia Icefield, *Geophysical Research Letters,* **25**(9):1407-1410.

Santana, A., 1984, Variación de las precipitaciones de 97 años en Punta Arenas como índice de posibles cambios climáticos, *Anales del Instituto de la Patagonia, Serie Ciencias Naturales,* **15**:51-60. (in Spanish).

Skvarca, P., Rott, H., and Nagler, T., 1995a, Synergy of ERS-1 SAR, X-SAR, Landsat TM imagery and aerial photography for glaciological studies of Viedma Glacier, southern Patagonia, Proceedings, VII Simposio Latinoamericano de Percepción Remota, SELPER, Puerto Vallarta, México, pp. 674-682.

Skvarca, P., Satow, K., Naruse, R., and Leiva, J. 1995b, Recent thinning, retreat and flow of Upsala Glacier, Patagonia, *Bulletin of Glacier Research,* **13**:11-20.

Skvarca, P., and Naruse, R., 1997, Dynamical behaviour of glaciar Perito Moreno, southern Patagonia, *Annals of Glaciology,* **24**:268-271.

Skvarca, P., Stuefer, M., and Rott, H., 1999, Temporal changes of Glaciar Mayo and Laguna Escondida, southern Patagonia, detected by remote sensing data, *Journal of Global and Planetary Change,* **22**(1-4):245-253.

Takeuchi, Y, Naruse, R., and Satow, K., 1995, Characteristics of heat balance and ablation on Moreno and Tyndall glaciers, Patagonia, in the summer 1993/94, *Bulletin of Glacier Research* (Japanese Society of Snow and Ice), **13**:45-56.

Warren, C., and Sugden, D., 1993, The Patagonian icefields: a glaciological review, *Arctic and Alpine Research,* **25**(4):316-331.

Warren, C., and Rivera, A., 1994, Non-linear climatic response of calving glaciers: a case study of Pío XI glacier, Chilean Patagonia, *Revista Chilena de Historia Natural,* **67**:385-394.

Warren, C., Rivera, A., and Post, A., 1997, Greatest Holocene advance of glaciar Pío XI, Chilean Patagonia: possible causes, *Annals of Glaciology,* 24:11-15.

Warrick, R. A., and Oerlemans, J., 1990, Sea level rise, in: *Climate Change: The IPCC Scientific Assessment,* J. T. Houghton, G. J. Jenkins and J. J. Ephraums, eds., Cambridge University Press, Cambridge, pp. 257-281.

THE SOUTHERN PATAGONIA ICEFIELD MASS BALANCE

An unsolved glaciological question

Juan Carlos Leiva[1*]

1. ABSTRACT

During the sessions of the IPCC[†] Working Group 1 – Scientific Assessment of Climate Change, in debates about the contribution of glaciers to rise in sea level, the relative contribution of the Patagonian icefields and Antarctica was considered to be unknown because of the lack mass balance data. Ten years later, with this problem still unsolved, this paper confronts the scientific community with the lack of data, in the hope that we might promote action to resolve the situation.

2. INTRODUCTION

Efforts made by the scientific community to assess how the climate change would affect the mean sea-level rise have been handicapped by the existence of a number of uncertainties. Among these, have been the contribution of the Antarctic, the possible instability of the West Antarctic Ice Sheet and also the role and contribution of glaciers and small ice caps (Warrick and Oerlemans, 1990).

The largest "small ice cap" in South America and one of the least well known from a glaciological point of view, is the Southern Patagonia Icefield (SPI). Its mass balance is largely unknown and therefore very little is known about its response to the present climate fluctuations. Only the front fluctuations of several outlet glaciers, their area shrinkage and a few ice-thinning rates are known.

[1] IANIGLA, Instituto Argentino de Nivología, Glaciología y Ciencias Ambientales, CC 330-5500, Mendoza, Argentina.

* corresponding author: jcleiva@lab.cricyt.edu.ar

† Intergovernmental Panel on Climate Change, organized by the World Meteorological Organization and the United Nations Environment Program.

The Patagonian Icefields: A Unique Natural Laboratory for Environmental and Climate Change Studies.
Edited by Gino Casassa et al., Kluwer Academic /Plenum Publishers, 2002.

3. THE MASS BALANCE CHALLENGE

The United States Geological Survey proposed a strategy for monitoring glaciers with three levels of surveillance (Fountain *et al.*, 1997):

- One glacier in each glacierized region is measured in detail to define the seasonal mass-balance processes, meteorological environment, and water runoff.
- Several glaciers in each region are measured to obtain only their annual mass balance.
- The changing areal extent and pattern of snow and ice exposure should be defined by remote sensing of glaciers using aerial photography and satellite images.

The scientific community is already aware that direct mass balance measurements cannot be used to assess the mass balance changes on regions as large as the Southern Patagonia Icefield, because they would be difficult and expensive to obtain. Therefore, the author proposes the selection of a few glaciers on each side of the icefield to perform these detailed and direct mass balance measurements and to use the remote sensing techniques to assess the mass balance for the whole icefield.

4. PAST AND PRESENT EXPECTATIONS OF THE GLACIER-MONITORING SCIENTIFIC COMMUNITY ON REMOTE SENSING

"Mapping, monitoring and inventorying the Earth's glaciers in the 21st century will continue to depend on data from remote sensors, both imaging and non-imaging, carried in aircraft and spacecraft, with increasing emphasis on the latter platform because of the global perspective and because of improvements in sensor capabilities." (Swithinbank, 1983).

"Non-imaging sensors, especially laser altimeters, with Global Positioning System (GPS) control, will provide accurate (sub-meter) profiles of glacier surfaces on a global basis. Not only will the topography of the Greenland and Antarctic ice sheets be measured consistently and accurately for the first time, but changes in surface elevation over time will be determined. When combined with information from imaging sensors about areal changes, changes in surface elevation will permit relative changes in volume to be determined." (Williams and Hall, 1998).

"The geodynamics laser altimeter system (GLAS) to be launched on an EOS polar-orbiting platform in about the year 2002, will provide high-resolution altimetric profiles of the ice sheets with a vertical accuracy of 5-10 cm. The altimeter mode will provide measurements of average surface elevation within a footprint of about 70 m in diameter. The primary application of these data will be to measure ice sheet rates of thickening or thinning and ultimately, to calculate changes in mass balance." (Williams and Hall, 1998).

If these wishes and hopes become true, the SPI mass balance challenge will be met. In the meantime, the glaciology group of the Icefields Scientific Task Force should forge the way to beginning mass balance research in the region.

5. REFERENCES

Fountain, A. G., Krimmel R. M., and Trabant D. C., 1997, *A Strategy for Monitoring Glaciers*, U.S. Geological Survey Circular, **1132**, United States Government Printing Office, pp. 19.

Swithinbank, C., 1983, Towards an inventory of the great ice sheets, *Geographiska Annaler*. **64A**(3-4):289-294.

Warrick, R., and Oerlemans J., 1990, Sea level rise, in: *Climate Change. The IPCC Scientific Assessment,* J. T. Houghton, G. J. Jenkins, and J. J. Ephraums, eds., Cambridge University Press, pp. 261-281.

Williams, R. S., and Hall D. K., 1998, Use of remote sensing techniques, in: *Into the Second Century of Worldwide Glacier Monitoring: Prospects and Strategies. A contribution to the IHP and the GEMS,* W. Haeberli, M. Hoelzle, and S. Suter, eds., Studies and Reports in Hydrology, **56**:97-111, UNESCO Publishing, France.

GLACIER MASS BALANCE AND CLIMATE IN THE SOUTH AMERICAN ANDES

An example from the tropics and a long-term and large scale concept for the Southern Patagonia Icefield

Georg Kaser[1*]

1. ABSTRACT

The spatial distribution of the energy balance and the rotation of the Earth cause zonal atmospheric circulation patterns. The South American Andes are an almost uninterrupted N-S stretching barrier to these main currents, reaching from the tropics into the trade winds, the westerlies and the subantarctic circulation patterns and causing more or less sharp climate contrasts. Consequently, glacier mass balance regimes vary within short E-W distances and the related glacier fluctuations provide a highly informative tool for studying present and past atmospheric circulation systems. First studies on this matter from the Cordillera Blanca, Perú, show the possibilities of climate-glacier studies in the Andes. In addition, the present paper proposes the design of a large scale and long-term mass balance program on the Southern Patagonia Icefield.

2. THE GLOBAL CIRCULATION OF THE ATMOSPHERE

Both the parallel incidence of solar radiation, and the radial emission of terrestrial radiation determine the spatial distribution of the energy budget on the Earth's surface. It is characterized by a net gain in the low latitudes and by a net loss at high latitudes. In a simplified view (Figure 1), this leads to a geopotential value of a given pressure level which is higher at low than at high latitudes. The resulting meridional pressure gradient is strongest where the energy budget changes from gain to loss. The consequent air movement is deviated by the rotation of the Earth (Coriolis deviation) resulting in the circumglobal westerlies where the frontal activities provide the atmospheric energy exchange between gain and loss regions.

---

[1] Institut für Geographie, University of Innsbruck, Innrain 52, A - 6020 Innsbruck, Austria.

* corresponding author: Georg.Kaser@uibk.ac.at

The Patagonian Icefields: A Unique Natural Laboratory for Environmental and Climate Change Studies.
Edited by Gino Casassa et al., Kluwer Academic /Plenum Publishers, 2002.

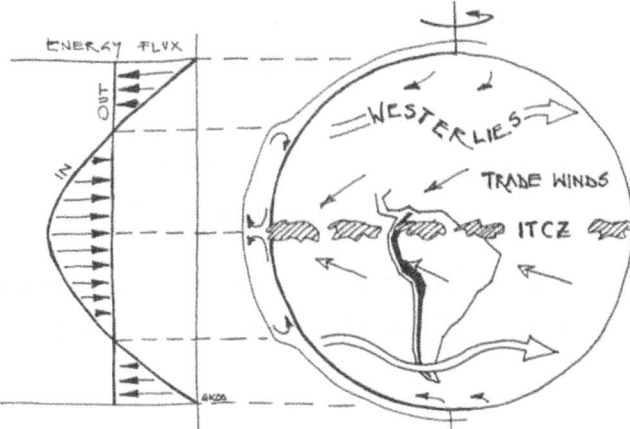

Figure 1. The global distribution of the net radiation fluxes causes, together with the rotation of the Earth, zonal patterns of the global atmospheric circulation. The South American Andes act as an almost uninterrupted barrier to the main currents.

Within the low latitude gain region, a two-cell Hadley circulation distributes the energy by air transport. There, the most marked patterns are the low level atmosphere trade winds and the Intertropical Convergence Zone, ITCZ.

Because of sinking effects, the trade winds are related to markedly dry conditions whereas the ITCZ is characterized by convective processes and humid conditions. On their way to lower latitudes, the trade winds are also initially deviated by the Coriolis force to the east and, consequently, also within the ITCZ the air masses basically move from east to west. The "outflow" of low atmosphere cold air from the polar regions is also deviated resulting in an east to west current. On a global view, the radiation geometry and the Earth's rotation account for the zonal character of the major circulation pattern of the atmosphere and each pattern stands for a typical climate.

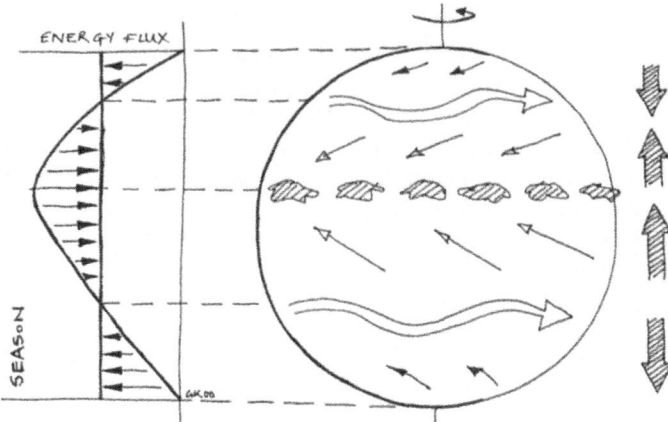

Figure 2. Idealized atmospheric circulation pattern corresponding to a Northern Hemisphere summer. The global circulation features are compressed in summer and expanded in winter, on each hemisphere.

The seasonality of the insolation, however, induces an oscillation of the ITCZ and, consequently compresses all zonal circulations in the respective summer hemisphere and expands them in the winter hemisphere (Figure 2). Thus, intermediate climates exist such as the Mediterranean and the outer tropical climate, where different regimes alternate during the course of the year.

In the case of any climatic change, caused by either terrestrial or extraterrestrial reasons, the extension of each region of energy gain or loss will change (Figure 3). Consequently, the zonal currents will be affected in their mean position, their oscillation, and their intensity. In turn, if any signs of a formerly different climate are found e.g. in the landscape, they can, if carefully chosen and interpreted, tell us about different global atmospheric circulation conditions. The South American Andes are very suitable for such investigations.

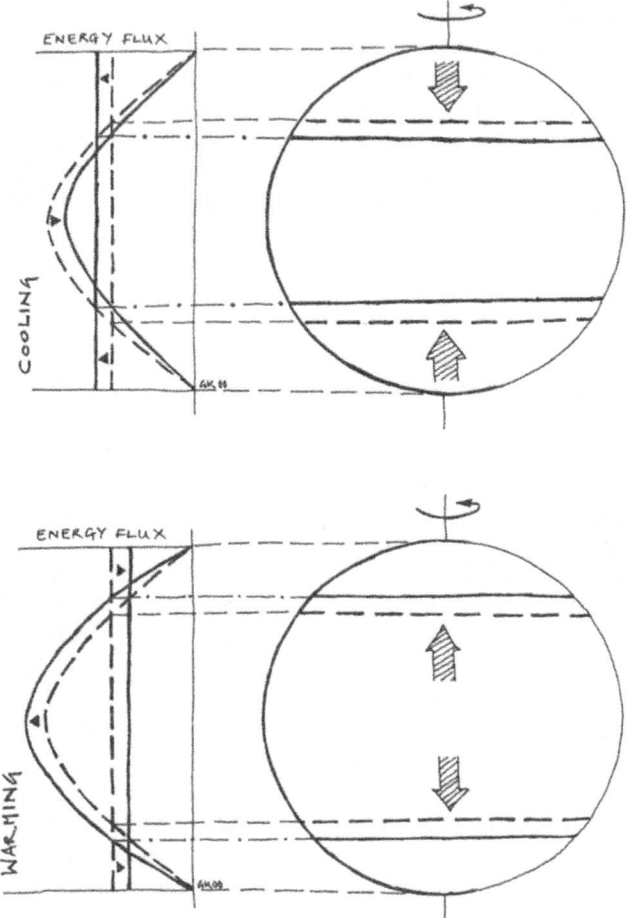

Figure 3. Changes in either the incoming or the outgoing energy fluxes cause a shift of the westerlies and thus compress (when cooling, top) or expand (when warming, bottom) all other zonal circulation features. Dashed lines indicate the original state and solid lines the perturbed state.

3. THE DIVERSITY ACROSS THE SOUTH AMERICAN ANDES

The South American Andes are an almost uninterrupted N-S stretching barrier to the zonal atmospheric circulations reaching from the tropics into the trade winds, the westerlies and the subantarctic circulation patterns and causing more or less marked windward and leeward climates. These will hardly be characterized by thermal differences since the temperatures do not vary substantially on a regional scale within one circulation system. But the atmospheric moisture content can vary highly within very short distances and this fact leads to sharp E-W gradients of all variables linked to it. In the low latitude climates of the ITCZ and the trade winds, the eastern slopes of the mountains are substantially wet whereas the western slopes are, in parts, extremely dry. This pattern changes clearly in the regions that are dominated by the westerlies. In the areas of seasonally changing regimes, the intensity of the gradients or even their sign may change within a year. If, however, circulation patterns and their seasonal oscillation were different in former climates, they should have left clues in the environment about their former character. The glaciers covering extended parts of the Andes are very sensitive climate indicators. The eventual records from past climates stored in ice cores and moraine deposits can help us to improve our knowledge about climate history substantially and in much detail, far beyond a simple temperature-precipitation discussion. The understanding of climate-glacier interactions is also essential for any water management plans in glaciated catchment areas from both the point of view of an eventual climate change and the obvious dramatic increase in water demand in many countries.

4. CLIMATE AND GLACIERS

The most immediate link between climate and glaciers is the glacier mass balance. Although they are not entirely independent of each other, precipitation, air temperature, atmospheric moisture and solar radiation are the major climate variables determining accumulation and ablation on a glacier. Other characteristics such as cloudiness, sublimation, and resublimation as well as the conditions of the glacier surface such as temperature and albedo depend on the former variables, whereas avalanches and calving processes are not immediately dependent on climate. Wind-induced drift may play a substantial role on both sides of the mass balance. The climate setting of a glacier is a combination of large scale influences and local effects, thus each glacier has to be "calibrated" as "climate-meter" before any climate information can be derived from observed or reconstructed glacier fluctuations.

The most significant key values for a glacier-climate investigation are the distribution of the specific mass balance with altitude or vertical balance profile, VBP, and the altitude of the equilibrium line, ELA. Both can be determined from field measurements for a respective period, say, one specific season or one year. Various more or less detailed models have been applied to describe the VBP under certain climate conditions and the sensitivity of the ELA to climate perturbations (e.g. Kuhn, 1980; Oerlemans, 2001). The latter effect depends on the VBP at the respective altitude range. If either the ELA or the volume of a former glacier can be determined from geomorphological evidence, the models allow us to estimate the climate that was responsible for the former glacier extent. This is normally restricted to large glacier extents since formerly small glaciers can hardly be reconstructed.

Local effects are usually found at topologically low portions of the glacier tongues due to shading or radiation emitted from surrounding slopes, but also leeward and windward sectors in the accumulation areas can substantially differ within one climate regime. Such effects can normally be compensated when looking at larger glaciated areas. If then, within one thermally homogeneous system, major differences in the glacier extent and the glacier behavior are found, conclusions can be made taking moisture-related climate parameters into account.

5. AN EXAMPLE FROM THE OUTER TROPICS

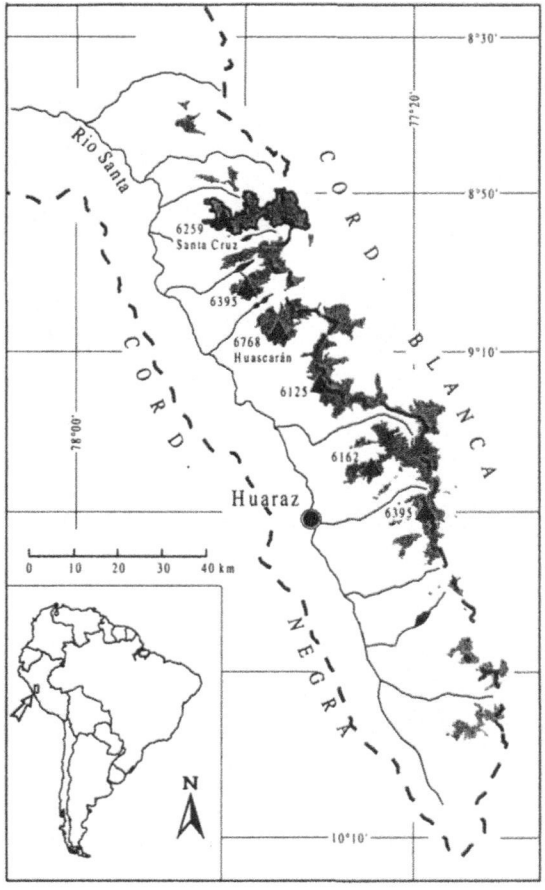

Figure 4. The Cordillera Blanca, Perú. Dashed lines indicate the watershed of the Rio Santa catchment basin; dotted areas indicate the glacier extension. A solid line surrounds the Santa Cruz - Alpamayo - Pucahirca glaciers. Triangles show the chain of the highest summits.

The Cordillera Blanca in Perú (Figure 4) is situated in the outer tropics, the intermediate zone between the all-year-round humid tropics and the dry subtropics. It faces tropical conditions under the influence of the ITCZ during the austral summer and subtropical conditions during winter (Kaser, 2001; Kaser and Osmaston, 2002). Under both conditions the advection of moisture is dominated by easterly air currents, thus the western slopes of the mountain range are markedly drier than the eastern slopes. Based on air photographs, taken in 1948 and 1950, we have analyzed two different extensions of the glaciers of the Santa Cruz–Alpamayo–Pucahirca group in the northern Cordillera Blanca (Georges, 1996; Kaser and Georges, 1997). Both extensions, the one derived from moraines formed in the 1920s and the one in 1950 were considered quasi stationary. The glaciers were grouped together into six different areas (Figure 5) and the different ELAs for each area and each glaciation were reconstructed by means of the hypsographic curves and assumed accumulation area ratios, AARs.

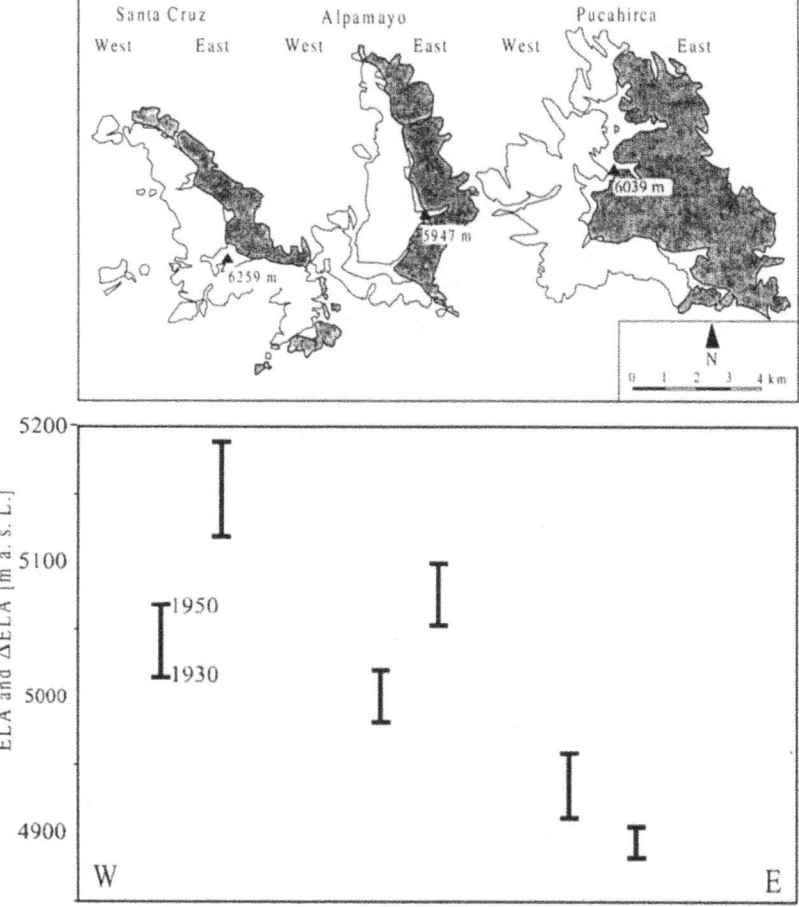

Figure 5. The six glacierized areas in the Santa Cruz - Alpamayo - Pucahirca range (top) and the distribution of the related equilibrium line altitudes, ELAs, in 1930 and 1950.

The obtained figure shows marked spatial differences in both the ELAs of each glaciation and the shift of the ELAs between 1930 and 1950. Both the general rise of the ELAs from east to west and the inverse situation on the two western mountains (not shown in Figure 5), can be explained by leeward and windward effects and, as a subsequent conclusion, a decrease in atmospheric moisture was found to be the major reason for the glacier retreat in the 1930s and 1940s.

Furthermore, the glacier-climate interaction in the outer tropics must be seen within the context of different regimes, the ITCZ-dominated inner tropical and the trade-wind dominated subtropical regimes, which alter from season to season. Each of these regimes produces different VBPs (Kaser and Georges, 1999; Kaser, 2001; Figure 6) which, in turn, cause different sensitivities of the ELAs to climate variables. Inner tropical glaciers are most sensitive to changes in temperature whereas subtropical glaciers are most sensitive to changes in atmospheric moisture (Kaser, 2001).

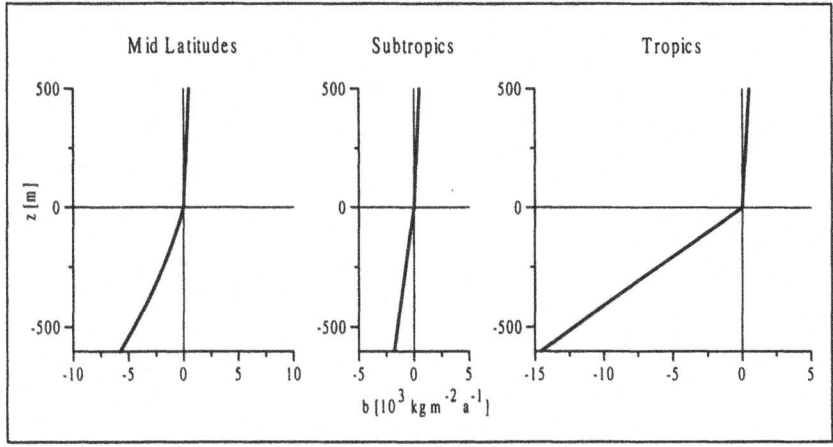

Figure 6. Typical profiles of the vertical balance profile, VBP, for different climate regions (from Kaser, 2001). The VBPs were modeled starting from a reference level z_{ref}, which is not identical with the equilibrium line altitude, ELA, but is the mean 0 °C elevation during the ablation period.

Both the duration and the intensity of each season control the fluctuations of the glaciers. Kull (1999) and Kull and Grosjean (2000) have shown that this concept is useful to investigate palaeocirculation patterns on former glaciations in northern Chile. In the Cordillera Blanca, in the year 2000, we started a detailed mass balance-climate study on two glaciers, with each representing one side of the mountain range. The analyses are in progress.

6. A CONCEPT FOR GLACIER MASS BALANCE STUDIES ON THE SOUTHERN PATAGONIA ICEFIELD

The Southern Patagonia Icefield, SPI (Figure 7) acts as a continental barrier against the atmospheric global circulation and provides strong zonal gradients of all moisture-related climate features such as precipitation, cloudiness, and latent heat flux.

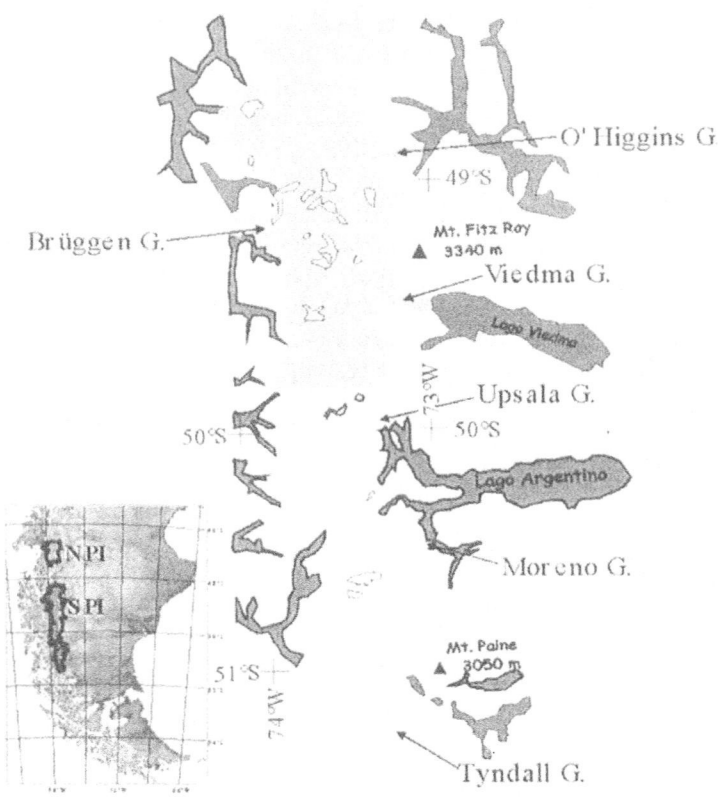

Figure 7. The Southern Patagonia Icefield with the main glaciers and proglacial lakes in the east (kindly made available by M. Stuefer).

Any study related to climate and climate change must therefore consider the context of the westerlies. The reconstruction of climate history, in turn, can teach us about former intensities and patterns of this feature of the global circulation and, thus, about the global circulation itself. From a more immediate point of view, the SPI provides an immense storage of water where the variations of runoff are strongly linked to the mass balance of the icefield and, in particular, the outlet-glaciers. A mass balance program, however, must be both designed for a long term investigation and be able to match the difficult logistic conditions which, apart from the large size of the SPI and its remoteness, include the unfavorable climate particularly on the windward side. Yet, the following information and data should be gathered at least along one or two east-west transects crossing the SPI (Figure 8):

- the horizontal long-distance profile of the specific mass balance
- on each side of the SPI, the vertical profile of the specific mass balance
- on each side of the SPI, the annual mean ELA

- on each side of the SPI, the glacier tongue variations
- climatological data (at least air temperature, atmospheric moisture, wind and global radiation) along the transect
- energy balance data from short periods on different sites along the transect
- the network also has to provide the ground truth for remote sensing data.

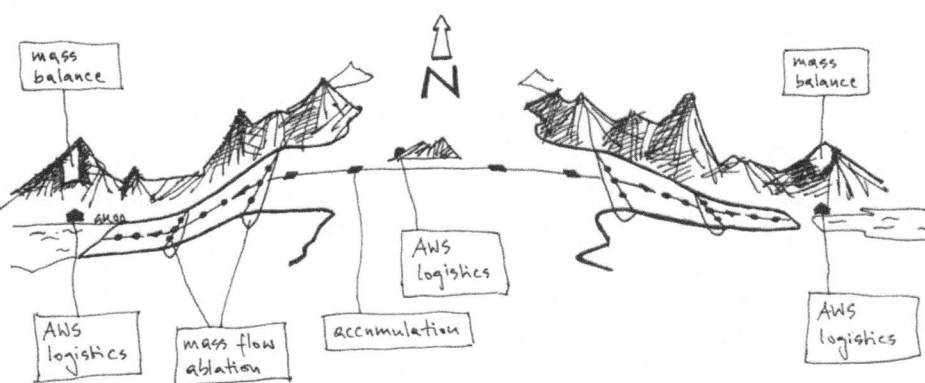

Figure 8. The concept of a possible future network to investigate glacier-climate interactions at the Southern Patagonia Icefield. AWS stands for Automatic Weather Station.

The mass balance has to be measured at selected points by ablation stakes and firn pits or cores. Accumulation measurements will be especially difficult, thus, the mass flow through a small number of "gates" must also be determined. The cross section is determined by geodetic and geophysical means, the velocity data are provided by the ablation stakes. The studies carried out recently by H. Rott and his team may represent a successful example (Rott *et al.*, 1998; Stuefer, 1999). If, possibly, a small mountain glacier is found on each side of the icefield, their VBPs and ELAs could inform us of the differences between the two sides of the icefield by comparatively easy means. The details about measurements and modeling of glacier mass balance are comprehensively discussed by a large number of different authors in Jansson *et al.* (1999). A strategy for monitoring glaciers is published by Fountain *et al.* (1997). The energy balance values can be extrapolated from the short-term measurements by means of climate data obtained from automatic weather stations (AWS).

Taking the logistic circumstances into account, the following network is proposed (Figure 8):

- Each transect should follow the central flow lines of two outlet glaciers being opposite to each other.
- The peak of the transect line should be close to an eventual drilling site.
- A cabin, possibly on a nunatak, could be the logistic point for any fieldwork on the top of the icefield including frequent accumulation measurements. This would also be the best site for an AWS.

- In front of each tongue, a fixed facility could provide support for an automatic weather station, AWS, and for any investigations in front of the tongue and on its lower portion, possibly including the monitoring of calving activities.
- Ablation and movement stakes should be placed along the flow lines and, at the "gates", across them.
- The choice of the transect(s) should be made considering their close proximity to a small mountain glacier on each side of the icefield.
- It is recommendable to select outlet glaciers with a hopefully simple geometry of the calving front(s).

7. SUMMARY

The zonality of the major global circulation systems leads to strong climate gradients and consequently, to different glacial regimes within short distances almost all over the South American Andes. These gradients are mainly caused by differences in variables that are linked to the atmospheric moisture, such as precipitation, cloudiness and evaporation. Glacier mass balance and climate studies provide a valuable tool for improving the understanding of the climate history in terms of detailed climate variables as well as in terms of fluctuations of the global circulation patterns. This has been successfully tested in the outer tropical Cordillera Blanca and in the subtropical Andes of northern Chile.

Any concept for a long-term mass balance program on the Southern Patagonia Icefield, SPI, must consider the difficult accessibility, the long distances and the inclement weather conditions.

8. ACKNOWLEDGMENTS

I am very grateful to the organizers of the Icefields Scientific Task Force on the *Aquiles* and in Valdivia in March 2000 for their extraordinary efforts and for their kind hospitality. I wish them every success in their efforts. I wish to thank Martin Stuefer, Christian Georges and Al Rasmussen for their careful reading of this manuscript.

9. REFERENCES

Fountain, A.G., Krimmel, R. M., and Trabant, D. C., 1997, A Strategy for Monitoring Glaciers, *U.S. Geological Survey Circular* 1132.

Georges, Ch., 1996, *Untersuchungen zu den rezenten Gletscherschwankungen in der nördlichen Cordillera Blanca*, Diploma thesis, University of Innsbruck, Austria (in German).

Jansson, P., Dyurgerov, M., Fountain, A., and Kaser, G., (eds.), 1999, *Methods of Mass Balance Measurements and Modelling*, Special issue of *Geografiska Annaler, 81A*(4).

Kaser, G., 2001, Glacier-climate interaction at low latitudes, *Journal of Glaciology*, 47(157):195-204.

Kaser, G., and Georges, Ch., 1997, Changes of the equilibrium line altitude in the tropical Cordillera Blanca (Perú) between 1930 and 1950 and their spatial variations, *Annals of Glaciology*, 24:344-349.

Kaser, G., and Georges, Ch., 1999, On the mass balance of low latitude glaciers, *Geografiska Annaler, 81A*(4):643-652.

Kaser, G., and Osmaston, H., 2002, *Tropical Glaciers*, International Hydrological Series, UNESCO, Cambridge University Press.

Kuhn, M., 1980, *Climate and Glaciers. Sea Level, Ice and Climatic Change* (Proceedings of the Canberra Symposium, December 1979) *IAHS Publications*, No. **131**:3-20.

Kull, C., 1999, Modellierung paläoklimatischer Verhältnisse basierend auf der jungpleistozänen Vergletscherung in Nordchile-Ein Fallbeispiel aus den Nordchilenischen Anden, *Zeitschrift für Gletscherkunde und Glazialgeologie*, **35**:35-64 (in German).

Kull, C., and Grosjean, M., 2000, Late Pleistocene climate conditions in the north Chilean Andes drawn from a climate-glacier model, *Journal of Glaciology*, **46**(155):622-632.

Oerlemans, J., 2001, *Glaciers and Climate Change*, Balkema Publishers, Rotterdam, Netherlands.

Rott, H., Stuefer, M., Siegel, A., Skvarca, P., and Eckstaller, A., 1998, Mass fluxes and dynamics of Moreno Glacier, Southern Patagonia Icefield, *Geophysical Research Letters*, **25**(9):1407-1410.

Stuefer, M., 1999, *Investigations on mass balance and dynamics of Moreno Glacier based on field measurements and satellite imagery,* Ph.D. thesis, University of Innsbruck, Austria.

ICE THICKNESS MEASUREMENTS ON THE SOUTHERN PATAGONIA ICEFIELD

Andrés Rivera[1, 2*] and Gino Casassa[2, 3]

1. ABSTRACT

The first detailed ice thickness measurements in the accumulation area of the Southern Patagonia Icefield have been obtained with a radio-echo sounding system, revealing a complex subglacial topography and internal reflection pattern.

A ground-based digital impulse radar system at 2.5 MHz was used to obtain continuous profiles of the subglacial topography. The system was mounted on sledges that were pulled by a snowmobile, allowing for coverage of extensive areas in short periods of time.

Two GPS receivers were used simultaneously and a differential correction method was applied for obtaining a precise geographic position at each thickness measurement.

2. INTRODUCTION

The Southern Patagonia Icefield (SPI), with an area of 13,000 km^2 (Aniya *et al.*, 1996), is the largest ice mass in the Southern Hemisphere outside of Antarctica. SPI has a length of 370 km (48-51° S) and an average width of 35 km (Figure 1), with 48 major outlet glaciers. Most of these glaciers have shown high retreat rates during the last decades, in response to the regional climate change (Aniya *et al.*, 1997). The glacier retreat is responsible for producing large amounts of meltwater which contributes significantly to the rise in sea level (Aniya, 1999). In spite of the generalized retreat, a few glaciers have shown strong advances during the last decades (Rivera *et al.*, 1997).

The observed changes of the SPI glaciers have been restricted mainly to the lower parts, the frontal and areal variations of the main glacier tongues. The accumulation area has been scarcely visited, due to logistic restrictions and harsh climatic conditions (Casassa *et al.*, 2002; Chapter 7). Ice thickness is one of the basic glaciological

[1] Facultad de Arquitectura y Urbanismo, Departmento de Geografía, Universidad de Chile, Marcoleta 250, Santiago, Chile; [2] Centro de Estudios Científicos (CECS), Arturo Prat 514, Valdivia, Chile; [3] Universidad de Magallanes, Casilla 113-D, Punta Arenas, Chile.

* corresponding author: arivera@uchile.cl

parameters needed for characterizing the glaciers. However, thickness data for the SPI are very limited, especially in the accumulation area.

This paper presents a review of the ice thickness methods and measurements carried out in Chile, as well as preliminary results of ice thickness and internal structure of the ice on the upper part of the SPI, by means of a radio-echo sounding system. The results are useful for a variety of scientific objectives, such as paleoclimate reconstruction, glacier modeling, and glacial geology, among others.

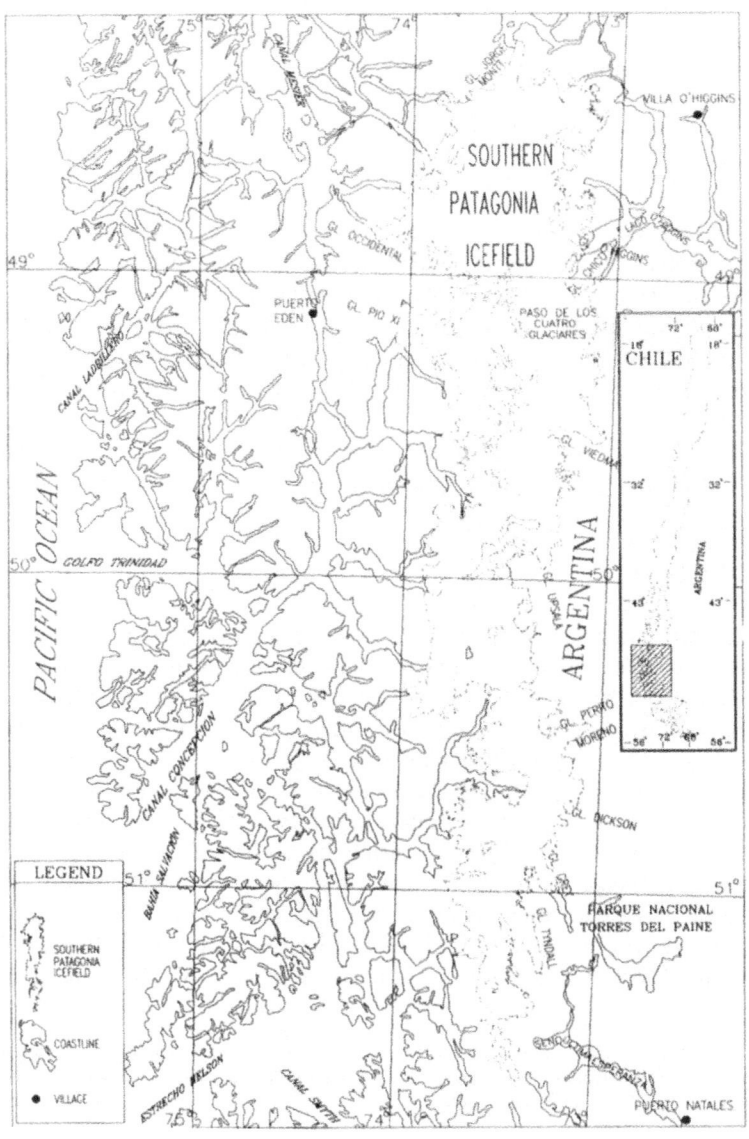

Figure 1. Location map of Southern Patagonia Icefield.

3. ICE THICKNESS METHODS

Two methods for detecting ice thickness have been applied in Chile:

3.1. Gravity Method

This method allows for determination of ice thickness based upon the deficiency of the vertical component of gravity observed on the surface of a glacier. This deficiency is generated by the smaller ice density with respect to the density of the underlying rock. Ice has a maximum density of 0.9 g/cm^3 (Paterson, 1994), while the density of the underlying rock is larger than 2.6 g/cm^3.

To convert the observed gravity to ice thickness values, it is first necessary to convert gravity data to a common reference value, such as mean sea level, obtaining in this way gravity anomalies (known as Bouguer anomalies) on ice. At the same time, it is necessary to measure gravity at rock outcrops surrounding the ice, to determine local and regional Bouguer anomalies due to varying geological conditions in the area. For example, intrusive rocks compose the "Patagonian Batholith", one of the most important units in the western margin of the SPI (Forsythe and Mpodozis, 1983). The differences between Bouguer anomalies on rock and ice result in residual anomalies due exclusively to the presence of the ice. Ice thickness is calculated applying an inverse model to the ice residual anomalies (Casassa, 1987).

The advantages of the gravity method are mainly logistic, due to the light weight of the equipment, little time required for each measurement and the unlimited maximum penetration range of the system. The limitation of this method is its great uncertainty (about 20% of the ice thickness) because the measured surface gravity value is in fact an areal average gravity of the underlying material for each station, which produces a smoothing of the complex subglacial topography.

3.2. Radio-Echo Sounding

Detection of ice thickness and internal structure of glaciers using radar, or radio-echo sounding, is also known as radioglaciology. This method is based on the same principle as sonar, frequently used in navigation and marine prospecting, and the seismic method, used in oil prospecting.

In essence, radio-echo sounding consists of the transmission of a signal into the glacier, which is reflected at the glacier bed, returning to the ice surface, where it is captured by a receiver. Ice thickness is determined based upon the so-called "two-way travel time", or time needed for the signal to travel from the transmitter to the glacier bed, and back to the receiver.

Different radio-echo sounding systems have been used to date, most of them working at frequencies between 30 and 700 MHz (Bogorodsky *et al.*, 1985). At this frequency range, cold ice is transparent to the penetration of electromagnetic waves. However, for temperate ice, that is ice at melting point, where water coexists with ice, radio signals with frequencies larger than 30 MHz are scattered by water bodies within the ice, and are also strongly attenuated by absorption, thus preventing wave penetration.

The first to suggest the construction of a radar for temperate ice were Watts and England (1976), who recommended frequencies lower than 10 MHz, where the scattering of the signal should theoretically be significantly reduced. Based on this principle,

Vickers and Bollen (1974) built the first radar for temperate ice, successfully probing the South Cascade glacier in Washington, USA. The transmitter was an impulse model type, generating a monopulse signal, which is transmitted using resistively-loaded dipole antennas.

At present, radio-echo sounding is widely used because it is the most accurate method (1-10%), which combines powerful data collection capabilities and field versatility.

4. RADIO-ECHO SOUNDING SYSTEMS USED IN CHILE

4.1. Analogue Radar

The first radar system used in Chile consisted of a transmitter with an amplitude of a few hundred Volts and an analogue receiver composed of a CRT oscilloscope, taking photographs of the screen for obtaining the data (Casassa, 1992). This system allowed for collection of discrete measurements.

4.2. Portable Digital Radar

To eliminate the technical problems arising from the need to photograph the oscilloscope screen, a second radar system was designed, composed of a transmitter with an output signal amplitude of 680 Vpp and a digital oscilloscope receiver with connection to a PC via RS232 interface (Casassa and Rivera, 1998). This system allowed for discrete measurements.

At present, the digital system is used, with antennas connected to fibreglass fishing poles of 8 m length, which are transported by two people, one at the receiver and one at the transmitter, using backpacks. This system allows for data collection while walking on the surface of the glacier, generating a continuous profile of subglacial returns. Due to its profiling nature, the method has improved the interpretation of subglacial returns and has been especially useful when applied to crevassed ablation areas, in high altitude glaciers, or accumulation areas with limited access (Rivera *et al.*, 2000a).

4.3. Snowmobile-Based Digital Profiling Radar

To measure ice thickness in extensive areas, a profiling system was designed, which allows for collection of ice thickness data every two seconds, creating a continuous profile of subglacial returns (Casassa *et al.*, 1998a).

In this profiling system, the transmitter and the digital receiver are mounted on sledges, pulled by a snowmobile. A 20 m half-dipole antenna length has been used, with the receiver and the transmitter being located 40 m and 100 m respectively behind the snowmobile. In this system the separation between the receiver and transmitter is 60 m (Figure 2).

The antennas are connected to an impulse transmitter with an output amplitude signal of 1600 Vpp. In the receiver, the antennas are connected through a balun to a digital oscilloscope. The data are recorded on the hard disk of a notebook computer. The receiver and transmitter systems are powered by 12 V batteries.

Ice thickness measurements are positioned with topographic data collected by either geodetic or topographic-quality GPS (Global Positioning System) receivers, using a differential correction method, which permits fixing positions every five seconds. In the portable system, a GPS receiver is installed on the backpack, whilst in the profiling system, a GPS receiver is mounted on the snowmobile (Casassa *et al.,* 1998b). The precision of the measurements depends on the GPS signal quality, which can reach sub-meter accuracy when geodetic-quality GPS receivers are used.

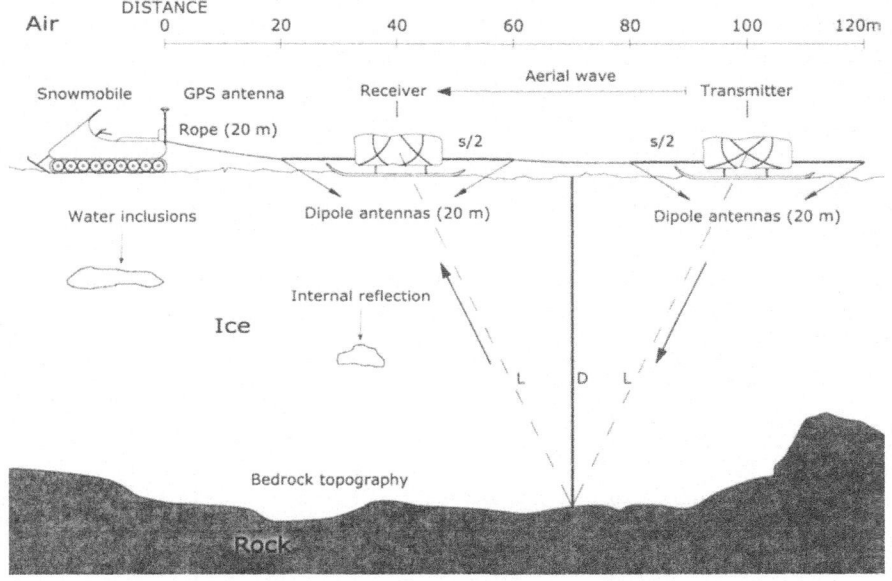

Figure 2. Snowmobile-based radio-echo sounding system. S/2 = 20 m is the half-dipole length of the antennas. The upper scale shows horizontal distance.

4.4. Helicopter-Borne Radar

Helicopter-borne and airplane-borne radio-echo sounding systems have been developed in Scandinavia, Russia, Germany, U.K., Italy, Japan, and the USA, among other countries. These systems have the advantage of rapid and effective measurement of remote glacier areas.

A Norwegian helicopter-borne radar has been used successfully in Svaritsen (Kennet *et al.,* 1993), consisting of an impulse transmitter with half-dipoles of 8 m and a central frequency of 6 MHz which can measure a thickness of 300 m of temperate ice. A similar system for temperate ice is being built at the University of Magallanes, Chile.

In polar ice, a 150 MHz German radar system has been used successfully to probe 2000 m of ice in Antarctica. The antenna is a corner reflector, consisting of two plane reflectors mounted at an angle of 90 degrees on the antenna frame, which is made of fiberglass tubes, all of which hangs from a helicopter. This system has been used recently in Chile (Damm *et al.*, 1999). Results show that its performance in temperate ice of Patagonia is very poor, compared to the good results obtained on clean ice in Antarctica and in the Alps.

4.5. Transmitters

Four different transmitters have been used in the radar systems of Universidad de Chile-Universidad de Magallanes, which are shown in Table 1.

The Bristol transmitter was designed by University of Bristol (UK), and has been used with success by Gilbert *et al.*, (1996) in southern Chile. The Narod/Clarke transmitter was designed by Barry Narod and Garry Clarke (Narod and Clarke, 1994), and tested by Cárdenas (1998) in Patagonia. The O.S.U. transmitter (Huffman, 1993) was designed by the Ohio State University, USA, and used successfully in cold and temperate glaciers by Thompson *et al.*, (1982) and Thompson *et al.* (1988). The Modified-O.S.U. transmitter was designed by Cárdenas (1998), nearly doubling the output voltage of the original O.S.U. transmitter, tested in Antarctica.

Table 1. Transmitters used by Universidad de Chile-Universidad de Magallanes.

Transmitter	Peak output power	Peak output voltage	Output resistive load	Rise Time	Consumption in Amperes	Pulse repetition rate	Dimensions	Weight
	(kW)	(V)	(Ω)	(ns)	(A)	(pps)	(mm³)	(kg)
Bristol	9	680	50	< 5	60 mA	?	13x13x12	0,7
Narod/Clarke	24	1100	50	< 2	180 mA	512	102x75x30	0,1
O.S.U.	51	1600	50	≈ 100	0,59 A	200/400/800	200x120x58	0,7
O.S.U. Modified	168	3000	50	≈ 100	0,8 A	200/400/800	200x120x58	0,7

4.6. Oscilloscopes

Different oscilloscopes have been used as part of the radar system of Universidad de Chile-Universidad de Magallanes:

- Hitachi V-209 analogue oscilloscope, with a bandwidth of 20 MHz
- Tektronics digital storage oscilloscope, model TekScope THS 720, with a 100 MHz bandwidth
- Fluke digital storage oscilloscope, model PM97, with a 50 MHz bandwidth
- Hitachi digital storage oscilloscope, model VC-6045A, with a bandwidth of 20 MHz
- Lecroy digital storage oscilloscope, model 9310-A, with a 400 MHz bandwidth.

5. ICE THICKNESS MEASUREMENTS IN CHILE

In the Antarctic Peninsula, the gravity method was applied by Chileans on Anvers Island ice cap in 1982 (Casassa, 1989) using precise measurements of gravity and station coordinates collected by German geodesists.

The first published measurements of ice thickness in Chile were carried out using the gravity method at the Northern Patagonia Icefield (NPI) by Casassa (1987), who estimated a maximum thickness of 1460 m on the accumulation area of San Quintín glacier, determining a subglacial topography under sea level.

In 1990, during the Japanese Glaciological Research Project in Patagonia (Naruse and Aniya, 1992), ice thickness at the ablation area of Tyndall Glacier (Figure 1) was measured with an analogue radio-echo sounding system (Casassa, 1992). The maximum penetration was 616 m near the center line at the ablation area of Tyndall Glacier. The high attenuation and absorption of the radar signals due to the presence of water inclusions in that part of the glacier, prevent the penetration of thicker ice. The maximum thickness of this part of the glacier could exceed 1000 m.

In 1992, a team of geologists from the University of Bristol, together with geologists from SERNAGEOMIN (National Mining and Geology Service, Chile) and a geographer of the University of Chile (Andrés Rivera), carried out measurements of ice thickness at the glacier located inside the caldera of Nevados de Sollipulli (38°59' S/71°31' W). The maximum ice thickness measured was 650 m, which was obtained using a combined system of radar sounding and gravity data (Gilbert *et al.*, 1996). The radar system was digital with PC connection through a serial port.

In 1993, the ablation area of Tyndall Glacier was measured again by a Japanese-Chilean team, but this time a digital radar system was used. The ice thickness results obtained were somewhat smaller than those of the previous 1990 campaign, presumably due to thinning of the ice during the three-year period (Casassa and Rivera, 1998).

In 1999 and 2000, a digital radio-echo sounding system was used by researchers from the University of Washington (Raymond *et al.*, 2000) to probe ca. 600 m thick ice at Tyndall Glacier, Patagonia. The radar system used is potentially capable of sounding ca. 1000 m of temperate ice.

In Patriot Hills, located at 80° S in the Antarctic interior, radar measurements have been carried out by Chilean expeditions in 1995, 1996 and 1997, using a system pulled by a snowmobile. The early measurements of 1995 were performed with a discrete digital radar system, and a maximum penetration of ca. 300 m (Cárdenas, 1998). In 1996 and 1997, a profiling radar system was used and an ice thickness of up to 1320 m of cold ice could be detected at Patriot Hills (Rivera *et al.*, 1998; Casassa *et al.*, 1998a and 1998b). In the vicinity of Patriot Hills, seismic studies indicated a maximum ice thickness of approximately 2 km.

In central Chile, most of the glaciers are smaller than those in Patagonia, but the surface topography is more complex and generally crevassed. For this reason, it is not possible to use a snowmobile profiling system. Instead, we have used a profiling system mounted on 8 m fiberglass poles and transported on backpacks by two people walking on the ice surface. This system was successfully used at the glacier of cerro Tapado (Rivera *et al.*, 2000a), Juncal Norte, San Francisco, and cerro Plomo. In all these cases, penetration was successful, allowing detection of the full subglacial topography. A maximum ice thickness of 225 m was detected at the ablation area of Juncal Norte Glacier (Rivera *et al.*, 2000b).

In 1999 the first helicopter-borne measurements of ice thickness were performed at Tyndall and Dickson Glaciers in Torres del Paine National Park, in addition to several high altitude glaciers of central Chile (Figure 1). Preliminary results are presented by Damm *et al.*, 1999.

6. RADIO-ECHO SOUNDING MEASUREMENTS IN PASO DE LOS CUATRO GLACIARES, SOUTHERN PATAGONIA ICEFIELD

During the Hielo Azul III operation of the Chilean Air Force (FACH) of 1997, we carried out several ice thickness measurements on the upper part of the SPI. These are the first measurements published for the accumulation area of SPI (Rivera and Casassa, 2000).

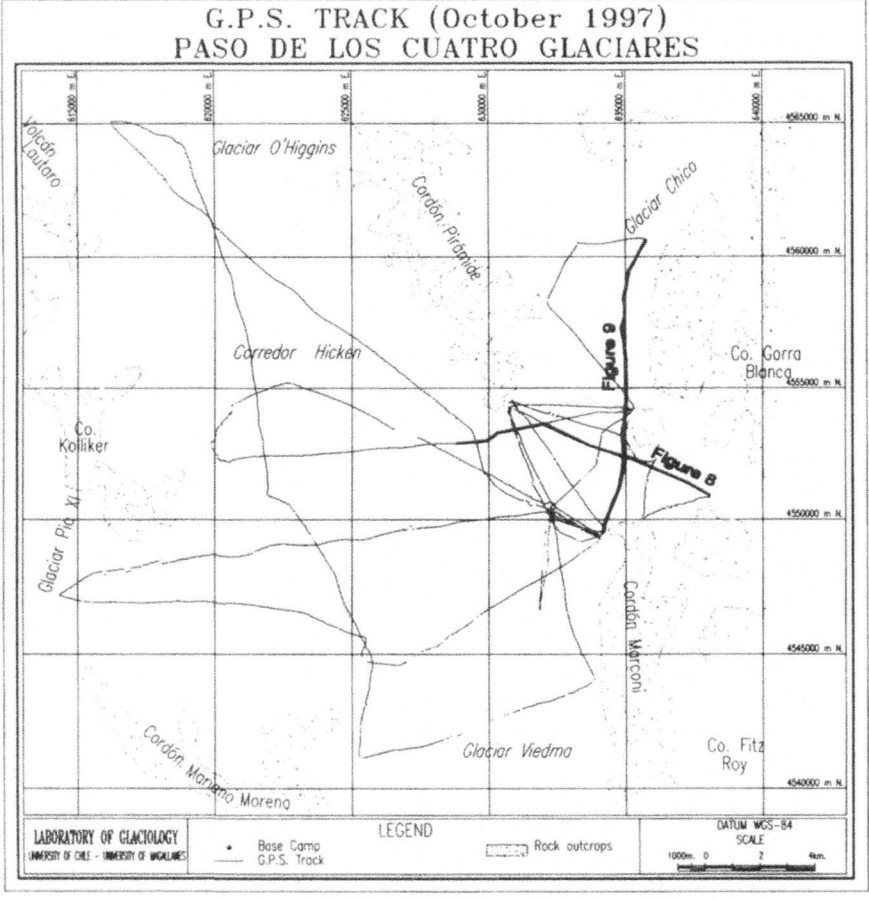

Figure 3. GPS tracks (dotted lines) on paso de los Cuatro Glaciares. The heavy lines correspond to the topographic profiles of Figures 8 and 9.

During the field campaign, base camp was located at 1450 m above sea level on the accumulation area of Chico Glacier, one of the outlet glaciers of the paso de los Cuatro Glaciares (Figures 1 and 3).

From this plateau, several glacier tongues discharge ice in different directions: Viedma Glacier flows toward the south; Chico Glacier flows to the northeast; O'Higgins Glacier flows to the north; Pío XI Glacier flows toward the west and Marconi Glacier flows to the east. This large area extends from the vicinities of monte Fitz Roy to the south to volcán Lautaro to the north.

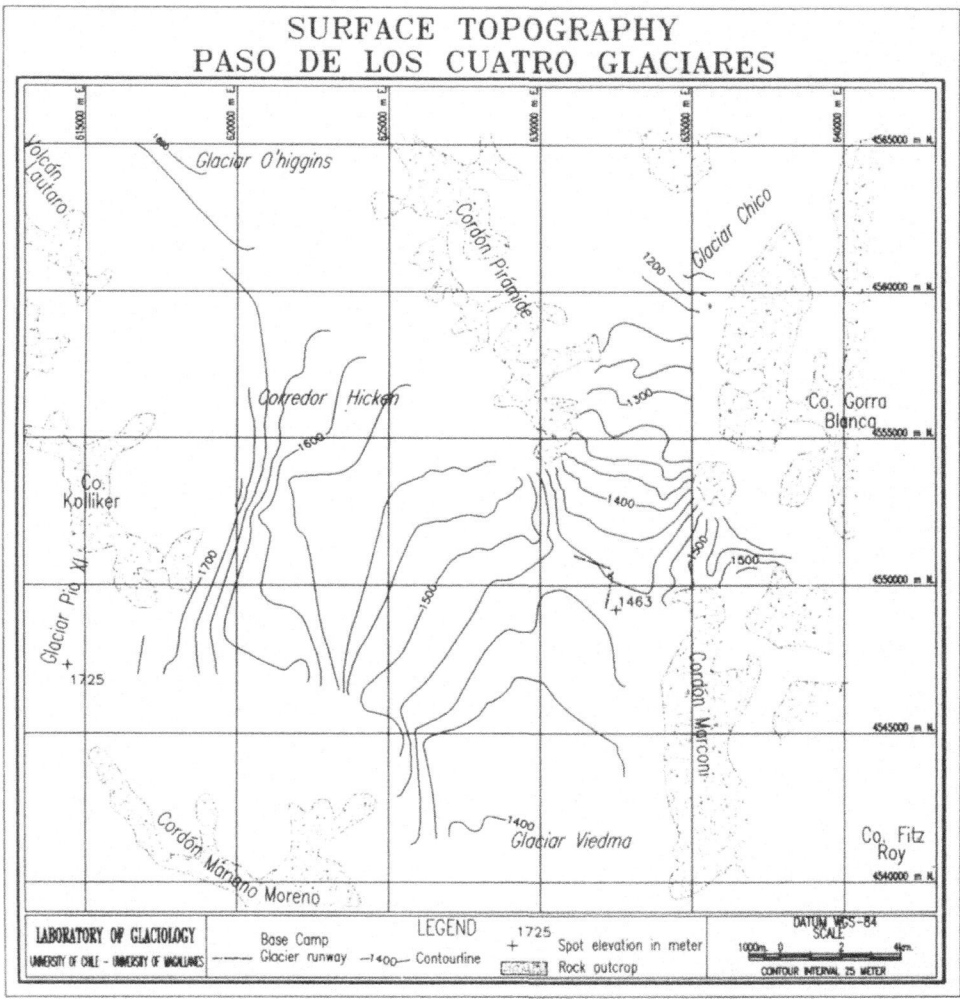

Figure 4. Surface topography of paso de los Cuatro Glaciares. The map has been adapted from Carta Preliminar 1:250,000 of Instituto Geográfico Militar de Chile. Elevation data collected in the field by the authors with GPS receivers have been used to compile contour lines at an interval of 25 m.

The recent behavior of glaciers of paso de los Cuatro Glaciares is contrasting. For example, Pío XI Glacier advanced 9 km in 50 years (Rivera *et al.*, 1997), O'Higgins Glacier retreated 15 km in 100 years (Casassa *et al.*, 1997) and Chico and Viedma Glaciers have retreated at slower rates (Rivera and Casassa, 2000 and Aniya *et al.*, 1997, respectively).

In the paso de los Cuatro Glaciares area, more than 25 radar profiles were measured (Figure 3). Each profile contains ice thickness measurements taken every two seconds.

To locate each measurement, we used one GPS receiver mounted on the snowmobile, which allows recording of one position every 5 seconds. Another GPS receiver was installed simultaneously at base camp, applying later a differential correction procedure to obtain surface topography with an accuracy better than 5 m on a horizontal scale and 10 m on a vertical scale.

Each ice thickness measurement was collected at the receiver and stored in real time by a PC, where a simultaneous log file with precise time was recorded. This information was necessary to combine the radar measurements with the GPS data, in order to locate each ice thickness data.

With the GPS position corrected, a surface topography map was produced (Figure 4). The contour lines have an interval of 25 m and their distribution shows a flat relief in the vicinity of the study area. The subglacial returns have been interpreted as the first significant and consistent increase of amplitude of the signal, after the end of the aerial wave. No migration or inversion procedures were applied to correct the bottom reflections.

By analyzing each radar profile with PC-based software, we could determine the subglacial topography for most of the profiles. In a few cases, the signal was too noisy, due to interference from the portable radio equipment used for personal communication. In other profiles, the signal was lost due to problems in the antenna connectors. Examples of ice thickness profiles are shown in Figures 5, 6, and 7.

Figure 5 shows the subglacial topography at the accumulation area of Chico Glacier from cordón Pirámide at the western side of the glacier (left) to cerro Gorra Blanca, and cordón Marconi at the eastern side of the glacier (right). In the central part of the main valley of Chico Glacier, deep ice not penetrated by our radar system can be observed. At the right side of this profile (cordón Marconi) it is possible to distinguish the rough subglacial topography, including one section where we believe a small subglacial lake may exist, at the site of a possible geologic fault.

Figure 6 shows another subglacial profile located at the accumulation area of Chico Glacier from cerro Gorra Blanca (left) to the Equilibrium Line Altitude (ELA) of Chico Glacier (right).

Figure 7 shows a radar profile close to volcán Lautaro at corredor Hicken. This profile contains two strong internal reflectors, at a depth of 296 m and 406 m. These reflectors have been interpreted as ash deposition layers produced by eruptions of the nearby volcán Lautaro, an active volcano. Its last eruption occurred in 1979 (Lliboutry, 1998) and several other eruptions have previously been reported during recent decades, in 1959, 1960, 1972 and 1978 (Martinic, 1988).

Figures 8 and 9 present two topographic profiles of the accumulation area of Chico Glacier, compiled partly from the radar profiles shown in Figures 5 and 6. The topographic profiles include both the subglacial and surface topography of the glacier. The location of both profiles is indicated in Figure 3.

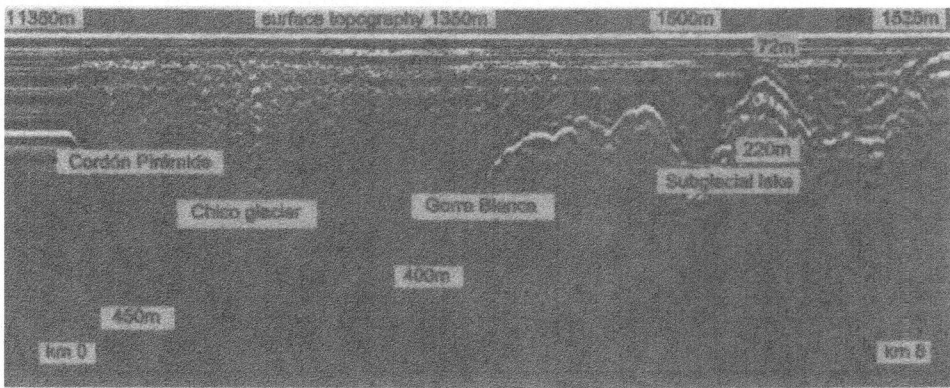

Figure 5. East-west radar profile of Chico Glacier. 72 m is the ice thickness above the subglacial nunatak. 220 m is the ice thickness above the feature interpreted as a subglacial lake.

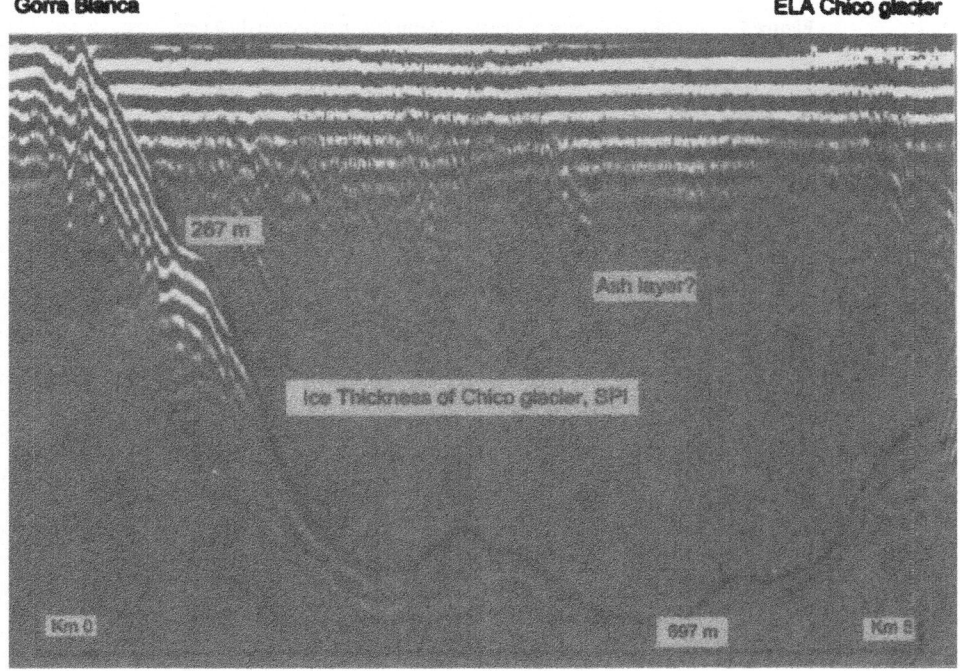

Figure 6. South-north radar profile of Chico Glacier. Numbers indicate ice thickness.

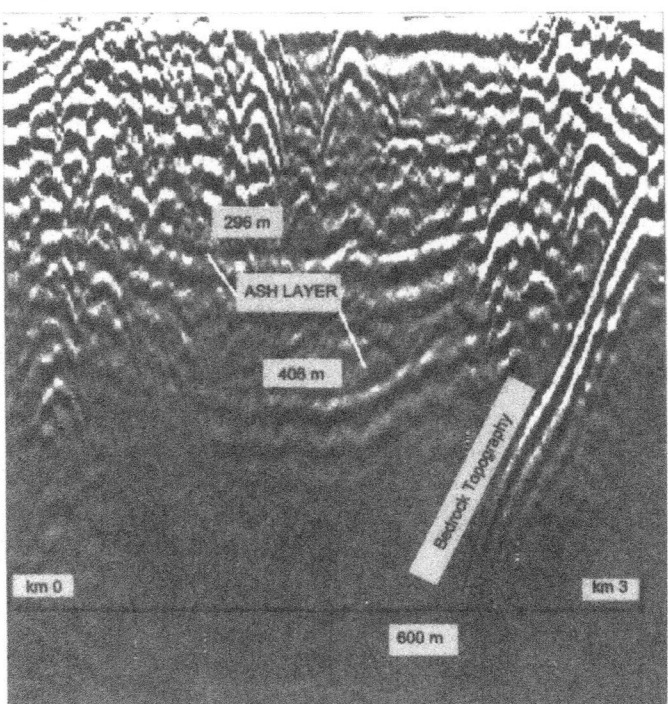

Figure 7. Radar profile at corredor Hicken. Numbers indicate ice thickness.

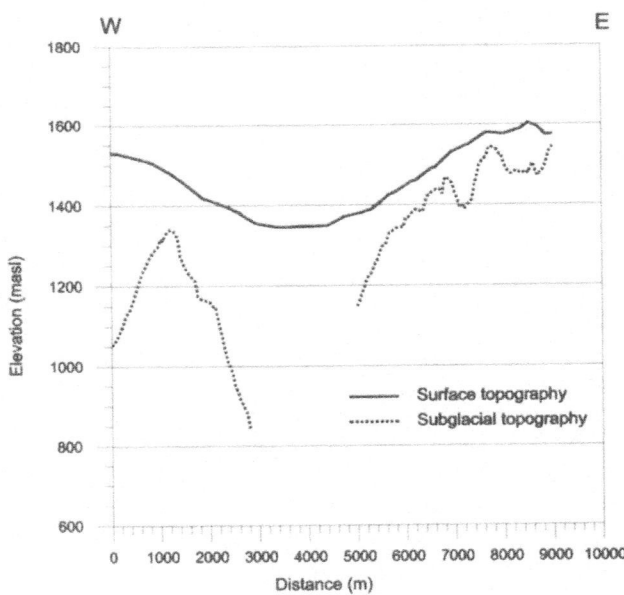

Figure 8. Topographic profile from west to east at the accumulation area of Chico Glacier. Note 7.5:1 vertical exaggeration.

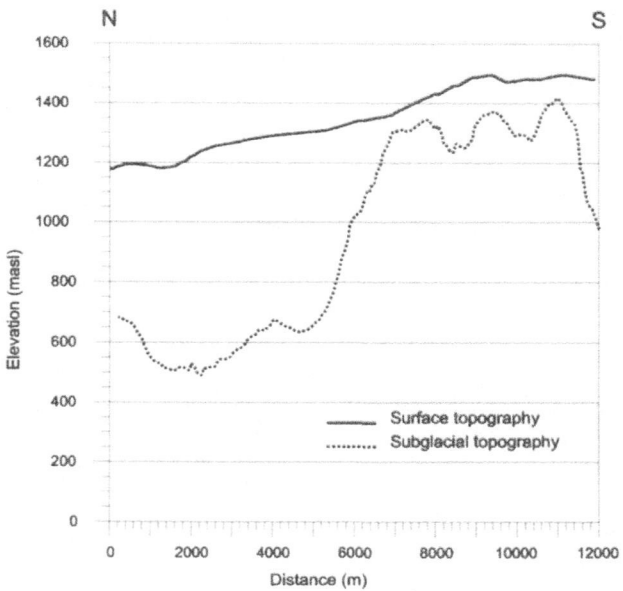

Figure 9. Topographic profile from north to south at Chico Glacier. Note 7.5:1 vertical exaggeration.

7. CONCLUSIONS

In spite of the progress made in ice thickness measurements over the last few years, it has been impossible to penetrate temperate ice deeper than 800 m in Patagonia. In this kind of temperate ice, the attenuation, absorption and scattering from water inclusions and internal reflectors, prevent collection of better results with the current system. Several modifications are being tested to improve the radar system range, such as increasing the amplitude of the transmitter signal, averaging a larger number of traces, and enhancing the signal strength by means of PC-based software.

The subglacial topography of paso de los Cuatro Glaciares cannot be completely detected with our system. Particularly difficult to probe, are the deep areas located in the central part of the main valleys. The interpretation of the subglacial returns could be improved by using a migration method, able to correct the returns from lateral reflectors, which are very important in areas where the bedrock topography is rough and steep, such as in Patagonia.

The internal structure of the ice needs to be better interpreted. At least two internal reflectors parallel to the surface topography have been detected at corredor Hicken. These reflectors suggest the presence of layers of volcanic ash deposited during one of the several eruptions of volcán Lautaro, located within the study area.

8. ACKNOWLEDGMENTS

This work has been carried out with financial support from project FONDECYT 1980293. The Chilean Air Force provided logistic support for the Hielo Azul operation.

Thanks are expressed to Cesar Acuña, for drawing the figures, Heiner Lange, for collaborating with the GPS data analyses and Carlos Cárdenas, for compiling Table 1. Comments from Al Rasmussen are acknowledged.

9. REFERENCES

Aniya, M., 1999, Recent glacier variations of the Hielos Patagónicos, South America, and their contribution to sea-level change, *Arctic and Alpine Research*, **31**(2):165-173.

Aniya, M., Sato, H., Naruse, R., Skvarca, P., and Casassa, G., 1996, The use of satellite and airborne imagery to inventory outlet glacier of the Southern Patagonia Icefield, South America, *Photogrammetric Engineering and Remote Sensing*, **62**:1361-1369.

Aniya, M., Sato, H., Naruse, R., Skvarca, P., and Casassa, G., 1997, Recent variations in the Southern Patagonia Icefield, South America, *Arctic and Alpine Research*, **29**:1-12.

Bogorodsky, V., Bentley C., and Gudmansen, P., 1985, *Radioglaciology*, D. Riedel Publishing Co., Netherlands, pp. 254.

Cárdenas, C., 1998, *Evaluación de transmisores de radar para hielo: aplicación en Patriot Hills, Antártica*, Undergraduate thesis, Civil Engineer, Universidad de Magallanes, Punta Arenas, Chile, pp. 213 (in Spanish).

Casassa, G., 1987, Ice thickness deduced from gravity anomalies on Soler Glacier, Nef Glacier and the Northern Patagonia Icefield, *Bulletin of Glacier Research*, **4**:43-57.

Casassa, G., 1989. Velocity, heat budget and mass balance at Anvers Island Ice Cap, Antarctic Peninsula, *Antarctic Record*, **33**(3):341-352.

Casassa, G., 1992, Radio-echo sounding of Tyndall Glacier, southern Patagonia, *Bulletin of Glacier Research*, **10**:69-74.

Casassa, G., Brecher, H., Rivera, A., and Aniya, M., 1997, A century-long recession record of Glacier O'Higgins, Chilean Patagonia, *Annals of Glaciology*, **24**:106-110.

Casassa, G., and Rivera, A., 1998, Digital radio-echo sounding at Tyndall Glacier, Patagonia, *Anales del Instituto de la Patagonia*, Serie Ciencias Naturales, **26**:129-135.

Casassa, G., Carvallo, R., Cárdenas, C., Jelinic, B., and Rivera, A., 1998a, Performance of a snowmobile-based radio-echo sounding system at Patriot Hills, Antarctica, *Proceedings VIII SCALOP Symposium*, Santiago, pp. 93-101.

Casassa, G., Rivera, A., Lange, H., Carvallo, R., Brecher, H., Cárdenas, C., and Smith, R., 1998b, Radar and GPS studies at Horseshoe Valley, Patriot Hills, Antarctica, *FRISP Report*, **12**:7-18.

Casassa, G., Rivera, A., Aniya, M., and Naruse, N., 2002, Current knowledge of the Southern Patagonia Icefield, in: *The Patagonian Icefields: a unique natural laboratory for environmental and climate change studies*, G. Casassa, F. V. Sepúlveda, and R. M. Sinclair, eds., Kluwer Academic/Plenum Publishers, New York, pp. 67-83.

Damm, V., Casassa, G., Eisenburger, D., and Jenett, M, 1999, Glaciological and hydrogeological studies of glaciers in central Chile and Patagonia using a helicopter borne radio-echo soundings system, Operational report and preliminary results, Archive-Nr. 0119119, Bundesanstalt für Geowissenschaften und Rohstoffe (BGR), Hannover, Germany, pp. 34.

Forsythe, R., and Mpodozis, C., 1983, Geología del basamento pre-Jurásico superior en el Archipiélago Madre de Dios, Magallanes, Chile, Boletín 39, Servicio Nacional de Geología y Minería, Santiago, pp. 63 (in Spanish).

Gilbert, J., Stasiuk, M., Lane, S., Adam, C., Murphy, M., Sparks, S., and Naranjo, J., 1996, Non-explosive, constructional evolution of the ice-filled caldera at volcán Sollipulli, Chile, *Bulletin of Volcanology*, **58**:67-83.

Huffman, F. E., 1993, Marx generator for high frequency ice radar system, Internal report, Department of Geological Sciences, The Ohio State University, Ohio, USA, pp. 26.

Kennet, M., Laumann, T., and Lund, C., 1993, Helicopter-borne radio-echo sounding of Svaritsen, Norway, *Annals of Glaciology*, **17**:23-26.

Lliboutry, L., 1998, Glaciers of Chile and Argentina, in: *Satellite Image Atlas of Glaciers of the World. Glaciers of South America.*, R. Williams and J. Ferrigno, eds., United States Geological Survey (USGS), Washington, 1386-I-6, I109-I206.

Martinic, M, 1988, Actividad volcánica histórica en la región de Magallanes, *Revista Geológica de Chile*, **15**(2):181-186 (in Spanish).

Narod, B. B., and Clarke, G. K., 1994, Miniature high-power impulse transmitter for radio-echo sounding, *Journal of Glaciology*, **40**(134):190-194.

Naruse, N., and Aniya, M., 1992, Outline of glacier research in Patagonia, 1990, *Bulletin of Glacier Research*, **10**:31-38.

Paterson, W., 1994, *The Physics of Glaciers*, 3rd edition, Pergamon Press, Great Britain, pp. 480.

Raymond, C., Neuman, T., Rignot, E., Rivera, A., and Casassa, G., 2000, Retreat of Tyndall Glaciers, Patagonia, Chile, *EOS, Transactions, American Geophysical Union*, **81**(48):F427, H61G-02.

Rivera, A., Aravena, J., and Casassa, G., 1997, Recent fluctuations of glaciar Pío XI, Patagonia: Discussion of a glacial surge hypothesis, *Mountain Research and Development*, **17**(4):309-322.

Rivera, A., Casassa, G., Carvallo, R., and Lange, H., 1998, Complex subglacial topography revealed under the Antarctic ice sheet at Patriot Hills, *Abstracts Antarctic Geodesy Symposium*, Universidad de Chile, Santiago.

Rivera, A., Giannini, A., Quinteros, J., and Schwikowski, M., 2000a, Ice thickness measurements on the glacier of Cerro Tapado, Norte Chico, Chile, Annual Report 1999, Labor für Radio- und Umweltchemie der Universität Bern und des Paul Scherrer Institut, Switzerland, Villigen: **38** (in German).

Rivera, A., Casassa, G., Acuña, C., and Lange, H., 2000b, Variaciones recientes de glaciares en Chile, *Revista Investigaciones Geográficas*, **34**:29-60 (in Spanish).

Rivera, A., and Casassa, G., 2000, Variaciones recientes y características de los glaciares Chico y O'Higgins, Campo de Hielo Sur, *IX Congreso Geológico Chileno, Actas*, **2**:244-248 (in Spanish).

Thompson, L., Bolzan, J., Brecher, H., Kruss, P., Mosley-Thompson, E., and Jezek, K., 1982, Geophysical investigations of the tropical Quelccaya Ice Cap, Peru, *Journal of Glaciology*, **28**(98):57-69.

Thompson, L., Xiaoling, W., Mosley-Thompson, E., and Zichu, X., 1988, Climatic records from the Dunde Ice Cap, China, *Annals of Glaciology*, **10**:178-182.

Vickers, R., and Bollen, R., 1974, An experiment in the radio-echo sounding of temperate glaciers, Menlo Park, California, Stanford Research Institute, Final report, pp. 16.

Watts, R., and England, A., 1976, Radio-echo sounding of temperate glaciers: ice properties and sounder design criteria, *Journal of Glaciology*, **17**(75):39-48.

LOCATING A DRILLING SITE ON THE PATAGONIAN ICEFIELDS

Niels Gundestrup[1*]

1. ABSTRACT

Locating a suitable drill site involves a number of compromises. At the proper place, an ice core can provide a record of climate changes in the past. Also, everything that has precipitated on the ice cap will be preserved, including biological fall out. At cold places, even gases are preserved in the bubbles in the ice, or in the ice lattice. In fact the ice cap can be considered as a long-term storage of both climatological and environmental information. In order to retrieve the information, an ice core has to be drilled. The site where the core is drilled determines the information that can be obtained. At a high accumulation site, it may be possible to obtain information at time scales of a fraction of a year. At a site with less accumulation, it may be possible to obtain records going further back in time. Summer melting obliterates the signals from the ice core, and for most studies summer melting should be minimized. The best drilling site therefore depends on the purpose of the drilling.

2. INTRODUCTION

Figure 1 shows an ideal ice cap with constant accumulation on a flat bedrock. The bedrock temperature is below the melting point, i.e. there is no basal sliding. The horizontal lines symbolize the snow that has fallen in one year. Also, lines showing the pattern followed by an ice particle, as well as a horizontal velocity profile are shown. It is assumed that the glacier is so cold, that no mass is lost due to evaporation or surface melt, and that the glacier is in steady state, i.e. that the snow accumulated on the surface is the same as the mass lost by melting and calving at the edge of the ice cap. Although this model is quite idealized, it shows some important features.

[1] Niels Gundestrup, University of Copenhagen, Juliane Maries Vej 30, DK-2100 Copenhagen, Denmark

* corresponding author: ng@gfy.ku.dk

The Patagonian Icefields: A Unique Natural Laboratory for Environmental and Climate Change Studies.
Edited by Gino Casassa et al., Kluwer Academic /Plenum Publishers, 2002.

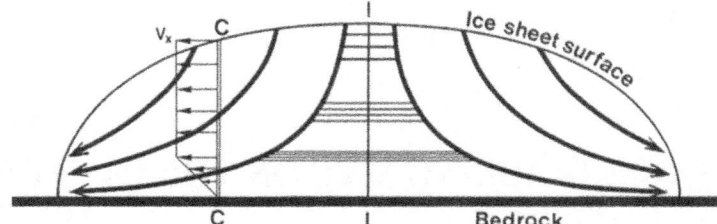

Figure 1. Simplified flow pattern and velocity of the ice in an ideal ice cap in steady state and with uniform accumulation. The figure also assumes no mass loss due to melting or evaporation and that the bottom temperature is below freezing.

Because the mass is preserved, the thinning of the layers with depth is compensated by a stretching of the layers. In the figure this means that the areas between the horizontal lines are constant with depth. If the accumulation is high, and the thickness of the glacier low, the annual layers will thin rapidly, limiting the time span recorded in the ice. For a typical glacier, the age of the layer halfway down in the ice cap will be approximately 0.6 x ice thickness/annual ice accumulation. This means, that if a site can be found with moderate accumulation, an ice core will reach far back in time. Cores with a length of about 300 m have provided records reaching back to the last glaciation, both in Canada and Greenland.

Figure 1 also shows, that if an ice core is not drilled at the top of the ice cap, the ice in the core will not originate from the drilling site, but from a site closer to the ice divide. Therefore, in order to interpret changes in the records from the ice core, the changes in the same records from the drilling site towards the summit must be known.

3. LOCATING A SITE IN THE SOUTHERN PATAGONIA ICEFIELDS (SPI)

In general, we must know the surface topography, ice temperature, bottom topography, spatial distribution of the different parameters and the internal layering in order to select a site where an ice core will provide a long undisturbed time record.

3.1. Surface Topography

A first requirement for selecting a drilling site is the *surface topography*. Because the ice always flows in the direction of the mean surface slope, and the velocity increases with surface slope, a good knowledge of the surface topography is required. Unfortunately, the surface is known with little accuracy. The reason for this is, that the photogrammetric technique normally used to map remote areas does not provide sufficient contrast on the interior of the ice cap to be reliable. One possibility might be to use satellite radar altimeters. However, due to the altitude of the satellite, and the relatively broad antenna pattern, the radar in the satellite will record the distance to nearest point to the satellite, and not the point directly below the satellite (Gundestrup *et. al.*, 1986). Although some of this error can be corrected for, the final accuracy will not be sufficient for our purposes.

Interferometric SAR (INSAR) from satellite or airplane may be a useful method. This technique requires the absolute 3-dimensional position of a few fixed points to be known, and it can then extend the measurements to provide a map. The technique requires two measurements with only little spacing between the antennae. By comparing the two images, elevation can be calculated. A large part of the globe has been mapped in this way, and considering the relatively small size of the SPI, this technique may be able to provide topography data. Again, sufficient changes in radar contrast are required. The INSAR technique is, however, unlikely to provide sufficient accuracy for determination of changes in elevation with time, and thereby changes in mass balance. The surface flow-velocities of the ice cap may also be determined in this way (Mohr *et al.*, 1998), provided that there is sufficient movement of the glacier surface before this surface has been changed by, for example, snowfall, melting, etc. The outlet of the Moreno Glacier in the SPI has been mapped using this technique (Rott *et al.*, 1998).

Several of the nations working in Greenland and Antarctica operate airplanes equipped with ice penetrating radars and GPS for positions.

NASA operates a 4-engine P3 airplane for remote sensing (Krabill *et al.*, 1995). The position of the airplane can be tracked to the dm level. It is equipped with a scanning laser altimeter and during work in Greenland, a 150 MHz radar sounder as well. This airplane would be able to map the entire SPI in a few days, and the radar sounder may be able to record ice thickness as well.

A more modest version of this method is operated by the Danish Cadastre (KMS). Here a laser distance meter is mounted in the bottom of a Twin Otter airplane (Figure 2). The airplane can at the same time be equipped with a radar ice sounder (Lintz Christensen *et al.*, 2000). Due to restrictions of antenna size, the frequency of the radar will have to be at least 35 MHz. This mapping could be performed by a Chilean Twin Otter. It will however require a significant amount of flight hours, because only a single profile is recorded.

Figure 2. Twin Otter airplane equipped with ice-penetrating radar. The 60 MHz monopole antenna is mounted in the tow hole through the tail section, minimizing the modifications required. At the bottom of the airplane a small laser distance meter is mounted, to measure the distance to the surface. On top of the airplane a dual frequency GPS antenna is mounted.

3.2. Accumulation and 10 m Ice Temperature

Knowing the topography, the next step will be to measure *accumulation* and *10 m ice temperature*. This is traditionally performed on 10 m ice cores (Paterson, 1994). However, due to the large accumulation at SPI, 20 m cores will be needed, for at least some of the sites. This should be supplemented with pole measurements on the most eastern part, where a pole can be located one year later. The pole measurements are needed, because the maritime climate, with small differences between winter and summer conditions, may mask the annual signal.

The unstable weather, and the high accumulation means that the 10 to 20 m cores should be drilled in a "hit and run" mode. Drilling equipment, generator and two operators will have a weight of about 500 kg. This will make it possible to have everything required for the coring in one Twin Otter or helicopter load. The ground time required is approximately 6 hours. Therefore, the airplane can stay on site for the time of the drilling. This will reduce the number of flight hours, and make it possible to terminate the drilling in case of approaching bad weather.

During the ground time, a GPS fix should be made. With a reference not more than 200 km away, it should be possible to obtain an accuracy of 10 cm, or significantly less than the height of the snowdrifts. These fix points can be used to check the more general, but less accurate maps. Also, the GPS positions will be the base for measurements of changes in mass balance, and thereby surface elevations.

Drilling an ice core also presents the possibility of recording the ice temperature at depths of 10 to 20 m. Low temperatures are required to obtain meaningful proxy data. At the Japanese/UMAG drilling at an elevation of 1750 m on Tyndall Glacier, water was found at 42 m depth (Godoi *et al.*, 2002; Chapter 14). Water will destroy most parameters that can be measured from the ice. Based on this drilling, a good drilling site should be at elevations well above 1750 m, hopefully at 2200 m or more.

Having learnt the surface topography, accumulation and 10 m ice temperature, the search for a drilling site can be more focused. Most likely, the promising area will be on the eastern part of the icefield, at elevations above 2200 m, and not too close to the ice margin. The accumulation will be one of the most critical parameters. Normally, the summit or ice divide will be the primary candidate for an ice-core drilling intended to reach far back in time. With the high accumulation at the SPI, it will more than likely be better to move further towards east, i.e. at the lee side of the ice cap. There are indications from the Argentine drilling (Aristarain and Delmas, 1993), that there is a very high west-east gradient in accumulation, although no reliable data are available for the icefield.

3.3. Bedrock Topography

Next, knowledge of the *bedrock topography* is required. The ice thickness has a significant impact on how many years an ice core will span. The bedrock is also important. If there are significant bedrock undulations, the layer thickness, and the sequence as well, may be destroyed for the ice close to bedrock, limiting the time span recorded in the ice. Test ice radar measurements have been made, showing thickness in excess of 750m (Rivera and Casassa, 2002; Chapter 10). For the SPI, the high temperature of the ice limits the frequency of the radar used. The 750 m recording was made with a 2.5 MHz monopulse radar. Also, a test with a 150 MHz radar has been performed (Damm and Casassa, 1999), but this radar could only penetrate the warm ice in a few locations, to a depth of approximately 700 m, at a measurement site in Torres del Paine National Park. A general description of the radar remote sensing technique is given in Bogorodsky *et al.* (1985). Although this book is rather old, it provides a good background for the technique. Seismic soundings have been performed at Moreno Glacier (Rott *et al.*, 1998), measuring a maximum of 720 m thick ice at a transect 8 km above the calving front. This technique requires however, strong logistic support on the ground.

It is essential to make the measurements from an airplane. The harsh climatic conditions on the SPI limit the possibilities for surface work, and it will be difficult to cover a larger area from the surface. Unfortunately, the only radar proven to work is the 2,5 MHz monopulse radar, and this radar cannot be used airborne due to antenna length restrictions. Although the test made with a 150 MHz radar was not successful, a higher frequency radar has a significantly larger dynamic range, and this higher dynamic range may be able to compensate for the higher loss in the warm ice. The higher frequency presents another problem: using a frequency where the wavelength in the ice is comparable to the layers in the ice, gives rise to multiple reflections in the ice, and if strong enough, they may mask the echo originating from the bedrock. As a result. the experiments with ice penetrating radars have to be continued. Due to the ice performance, it is not possible to identify a suitable radar based on specifications alone. With 35, 60 and 150 MHz commonly used for ice penetrating radars, there should be room for experimentation.

3.4. Ice-core Drilling

Knowing surface and bedrock topography, accumulation and 10 m ice temperature, it should be possible to limit the potential drilling sites to only a few. The next step is to drill an ice core to approximately 100 m. The reasons for requiring this depth are several: A 100 m ice core can be drilled in a few days, limiting the logistic problems (Figure 3).

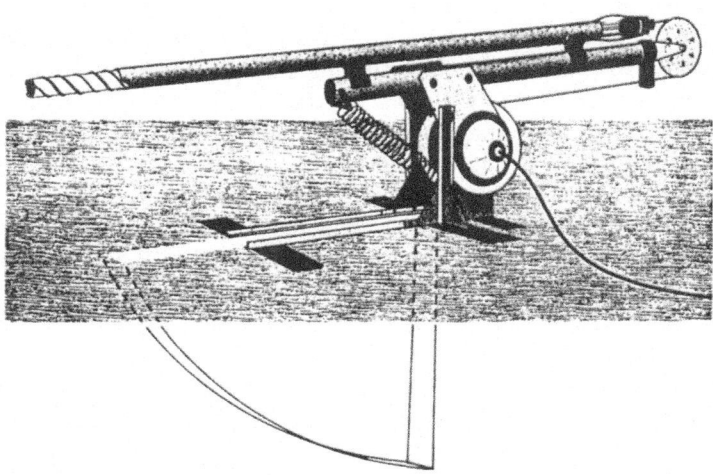

Figure 3. Shallow drill used by the University of Copenhagen. The drill is 2 m long. It is optimized for drilling 78 mm ice cores to depths between 20 and 140 m.

Also, at depths greater than 120 m, the core may start to fracture, although it is not known how severe this problem is for the ice at SPI. Finally, the open hole will decrease in diameter due to the overburden pressure from the ice, eventually closing the borehole around the drill. The closure rate depends heavily on the ice temperature, and it is unlikely that boreholes deeper than 200 m can be drilled open.

Deeper boreholes will require the hole to be liquid-filled in order to compensate for pressure from the ice. This requires a different drill and more complicated drilling procedures, resulting in significantly-increased logistic problems. It should be possible to drill to 500 m with a small camp in one season. Going deeper, a full-size deep drilling camp is required. This is a very expensive operation, and the climatic conditions at SPI make this operation difficult as well. As a result, the search for a good drilling site should concentrate on areas with an ice thickness of less than 500 m.

4. POSSIBLE SCHEDULE FOR DRILL-SITE LOCATION

The following is a very rough first attempt at producing a schedule for the drill-site reconnaissance. The actual schedule will naturally have to be changed depending on funding, logistic capabilities, etc. The purpose is to show how the various parts of the investigation depend on prior knowledge. It will be possible to accelerate the process, but especially because of its size, the SPI is unusually complicated because of lack of present knowledge, high accumulation and temperature, and adverse weather conditions.

- Step 1: Locate all the available information on surface topography. Next, locate areas with reasonable ice caps at elevations above 2200 m. The relevant areas should have a fair ice cap shape, and not be placed on a sloping mountain side. Pictures such as the NASA image shown in Benn and Evans (1998) are useful.
- Step 2: Measure the surface topography from an airplane. A few general lines should be flown connecting all of the SPI. Except for these few lines, the measurements will be concentrated according to the information from Step 1. It will be tempting to try one or more types of ice-penetrating radar on these flights. Based on the first flight results, the flight plan will be adjusted to provide a more dense coverage on the promising areas.
- Step 3: Knowing the surface topography, a number of 10 to 20 m-deep ice cores will be drilled in a "hit and run" technique. On the eastern side, this is supplemented with accumulation poles. The cores are selected so that they both give a general picture of the accumulation distribution, as well as a more detailed knowledge at the promising sites. The number of cores will most likely be 10 to 20. GPS will be used at all sites. These shallow cores are analyzed for stable isotopes, dust, ions and possibly organic material. They will be firn cores, as the depth will not be sufficient to reach solid ice.
- Step 4: If not done previously, ice-penetrating radars should be used to measure the ice thickness. The measurements could be concentrated on areas based on prior knowledge. Also, the flight plan can be changed to give improved coverage at areas with an ice thickness less than 600 m in the dry ice zone. This relatively shallow depth will increase the probability for the ice radar to give a good bedrock echo. The internal layering might occasionally show up on the radar returns. At this stage, all the critical information needed for the drill site

selection will be available. A complete review of the results should take place, and additional measurements may have to be made before it is possible to continue with the next step. In additional to the glaciological information, it will be important to include logistic information, i.e. weather condition at the drilling site, accessibility, etc.

- Step 5: Drill a 100 m ice core. The drilling site will most likely be on the eastern side of the ice cap, at an elevation of over 2200 m, in an area with not too high accumulation, and with an ice thickness of approximately 350 m. The drilling operation will take 3 days, and require, say, 2 tons of equipment. Again, depending on the result from this core, it may be necessary to go back and perform additional measurements. It is possible that more than one 100 m core may be required.

- Step 6: Core to bedrock. The core will have to be drilled in a liquid-filled borehole. A 350 m core can be drilled in 4 weeks. Due to the length of stay on the ice cap, and the number of persons involved, this operation will require extensive logistics. The total weight uplifted will be approximately 10 tons, including borehole liquid.

5. CONCLUSION

The Southern Patagonia Icefield is an extremely demanding site for icefield-related research due to the unstable weather conditions. However, using the techniques that are currently available, it would be possible to perform most of the site survey during good-weather days and without having to remain on the site for more than a few hours at a time.

6. REFERENCES

Aristarain, A. J., and Delmas, R. J., 1993, Firn-core study from the Southern Patagonia ice cap, South America, *Journal of Glaciology*, **39**(132):249-254.

Benn, D. I., and Evans, D. J. A., 1998, *Glaciers and Glaciation*, Arnold, ISBN 0 340 58431, Figure 1.18, Plate 4.

Bogorodsky, V. V., Bentley, C. R., and Gudmandsen, P. E., 1985, *Radioglaciology*, D. Riedel, ISBN 90-277-1893-8.

Damm, V., and Casassa, G., 1999, Glaciological and hydrogeological studies of glaciers in central Chile and Patagonia using a helicopter borne radio echo soundings system. Operational report and preliminary results, Archiv-Nr. 0119119, Bundesanstalt fur Geowissenschaften und Rohstoffe, Hannover, Germany, pp. 34.

Godoi, M. A., Shiraiwa, T., Kohshima, S., and Kubota, K., 2002, Firn-core drilling operation at Tyndall Glacier, Southern Patagonia Icefield, in: *The Patagonian Icefields: a unique natural laboratory for environmental and climate change studies*, G. Casassa, F. V. Sepúlveda, and R. M. Sinclair, eds., Kluwer Academic/Plenum Publishers, New York, pp. 149-156.

Gundestrup, N. S., Bindschadler, R. A., and Zwally, H. J., 1986, Seasat range measurements verified on a 3-D ice sheet, *Annals of Glaciology*, **8**:69-72.

Krabill, W., Thomas, R., Martin, C., Swift, R., and Fredrick, E., 1995, Accuracy of laser altimetry over the Greenland Ice Sheet., *International Journal of Remote Sensing*, **16**(7): 1221-1222.

Lintz Christensen, E., Reeh, N., Forsberg, R., Hjelm Jørgensen, R., Skov N., and Woelders K., 2000, *Journal of Glaciology*, **46**(154):531-537.

Mohr, J. J., Reeh, N., and Madsen, S., 1998, Three-dimensional glacial flow and surface elevation measured with radar interferometry, *Nature*, **391**:273-276.

Paterson, W. S. B., 1994, *The Physics of Glaciers*, 3rd edition, Elsevier, ISBN 0-08-037945.

Rivera, A., and Casassa G., 2002, Ice thickness measurements on the Southern Patagonia Icefield, in: *The Patagonian Icefields: a unique natural laboratory for environmental and climate change studies*, G. Casassa, F. V. Sepúlveda, and R. M. Sinclair, eds., Kluwer Academic/Plenum Publishers, New York, pp. 101-115.

Rott, H., Stuefer, M., Siegel, A., Skvarca, P., and Eckstaller, A., 1998, Mass fluxes and dynamics of Moreno Glacier, Southern Patagonia Icefield, *Geophysical Research Letters*, **25(9)**:1407-1410.

ON THE POTENTIAL TO RETRIEVE CLIMATIC AND ENVIRONMENTAL INFORMATION FROM ICE-CORE SITES SUFFERING PERIODICAL MELT, WITH SPECIFIC ASSESSMENT OF THE SOUTHERN PATAGONIA ICEFIELD

Veijo Pohjola[1*]

1. ABSTRACT

Problems related to the preservation of atmospherically-deposited signals in an icefield are presented and their implication for a potential ice-core record is briefly discussed. The largest problems regarding ice cores from warm and wet icefields is the washout and re-deposition of the signal during percolation and refreeze of meltwater. Just how much this process alters the signal, can be assessed by making a set of diagnostics on an ice-core record. An example how this can be done is given from an Arctic icefield. Finally the potential to retrieve a good ice-core record from the Southern Patagonia Icefield is discussed. Based on earlier work it seems likely that a good record can be retrieved on the highest and driest areas of the icefield.

2. INTRODUCTION

Since Langway's pioneering work on retrieving climatic data from ice cores in the 1950's (Langway, 1967) the efforts to extract information from these frozen archives have grown almost exponentially with time. In a perfect world, the snowfield and its packed strata of ice beneath its snow surface serves as a superb frozen archive of the chemical and physical properties of precipitating air masses. Getting the record of these properties gives us an opportunity to retrieve past climatic and environmental changes, either from a static view, by logging the changes over the snowfield, or from a dynamic view, by tagging the properties to different types of air masses, and resolving the interplay of those air masses to get a geographically wider perspective when changes can

[1] Department of Earth Sciences, Uppsala University, Villavägen 16, 752 36 Uppsala, Sweden.

* corresponding author: veijo.pohjola@geo.uu.se

The Patagonian Icefields: A Unique Natural Laboratory for Environmental and Climate Change Studies.
Edited by Gino Casassa et al., Kluwer Academic /Plenum Publishers, 2002.

be traced to the area of origin of, and the travel paths for, the statistically-dominant air masses.

In order to obtain a perfect-world situation, we need to know that the values of the properties we measure are proportional to the precipitated signal, that no post-depositional processes have affected the record, and that the signal is layered in perfect sedimentary strata season by season. The perfect world seldom appears in reality, though the closest approximation would be in a cold snow field, where temperatures in the snow pack never reach melt temperatures, and where precipitation is large enough to even-out stochastic effects of precipitation and wind drift. This means that a narrow search window exists for perfect sites to drill for an ice core, since temperature and precipitation are proportional to each other, due to the relation with vapor pressure in the air, meaning cold sites are dry, and wet sites are warm. Thus sites that are cold enough will be harassed by uneven snow deposition and wind drift effects, while sites with larger precipitation, though having less problems with uneven snow deposition may be affected by temporal melt in the snow pack.

Traditionally the temperature/melt problem has been viewed as a larger problem than the effect of small accumulation rates, which have promoted cold dry icefields in central Greenland and Antarctica for ice-core investigations. However, by investigating the record from these polar areas from both a static and a dynamic viewpoint, the outcome will be restricted to long-time global effects and to regional effects around these two vast and very continentally-influenced snow fields. In order to understand shorter time-scale changes of more temperate and tropical regions, we need ice-core records closer to these regions, which carry a larger maritime signal, and that are also wetter and warmer.

The largest temperate icefield in the Southern Hemisphere outside of Antarctica is the Southern Patagonia Icefield (SPI). This ice sheet is in close proximity to the southern Pacific Ocean, and therefore very wet and warm. Its highest mountains rise above 3000 m asl, which gives the potential for good conditions to preserve atmospherically deposited signals in the ice strata. At the Icefields Workshop held on the Chilean navy vessel *Aquiles* in March 2000, the potential for retrieving a good ice-core record from the SPI was discussed. This paper focuses first on the general problems of the distortion between the atmospheric signals and an ice-core record and then shows ways to analyze the effect of distortion by percolating water. This is managed by showing examples from work done on another maritime icefield, the Lomonosovfonna icefield, from the Svalbard archipelago in the Arctic part of the Atlantic Ocean. Finally, this paper discusses the potential to retrieve good atmospheric records from the SPI.

3. PROBLEMS RELATED TO THE RETRIEVAL OF CLIMATE SIGNALS FROM AN ICE CORE

Before approaching the problems of retrieving signals, the signals themselves need to be defined. The atmospherically-deposited signals we focus on in this work are the isotopic and chemical signal from the precipitated water, as well as the physical signals of ice structures found in the ice column. The signals from organic matter and substances, as well as the particulate matter of dust will not be dealt with.

Problems with finding the signal can be grouped within depositional and post-depositional effects. Other problems are related to sampling and analytic methods, which are not mentioned here. The largest depositional problem is the need for an even deposition of the signal during the period over which the signal is integrated (sample size). If no deposition has occurred during part of this interval, this time interval has no record, and time resolution is lost. In macro perspective this is related to storm tracks over the icefield/sheet and the accumulation pattern steered by local topography. The precipitation rate may also be seasonally varying, which biases the ice pack favorably to one, or several seasons (Cuffey and Stieg, 1998; Noone et al., 1999; Werner et al., 2000). In closer resolution the accumulation pattern is also related to turbulence and wind drift, that causes sastrugis, snow ripples, and erosion of older strata, which will mix older signals with younger ones (Goodwin et al., 1994; Stenberg, 2000). The wind drift is also operational on the post-depositional side of effects.

Other post-depositional effects are diffusion of the signal due to vapor transport in the ice pack, transposition of the signal due to water transport and chemical reactions in the ice pack. All these effects are very dependent on the temperature and density within the ice pack. The higher the temperature, the higher the pore volume in the ice pack, and the thinner the stratigraphic layers, the more important these processes become. The vapor transport in the ice pack acts as a smoothing filter due to diffusion of the deposited signals. This is also called maturing of an ice pack since small time scale variations will disappear, and the stronger spectrum of signals will be preserved (as a seasonal cycle). The diffusion of water isotopes is relatively well known and can be calculated (Whillans and Grootes, 1985; Johnsen et al., 2000), while the diffusion of chemical signals is less well investigated. The vapor-driven diffusion of a chemical signal would probably be proportional to each species' solubility in water, as in the potential to turn into the gas phase. The diffusion of chemical signals is also related to the problem with chemical reactions in the ice pack that alters the chemical loading (Legrand and Mayewski, 1997). This field of research is still young. Examples of processes that chemically alter a firn pack have been given by Nutterli et al. (1999) and by Pasteur and Mulvaney (1999).

The transposition and smoothing of the signal is further due to infiltration and percolation of water related to the melting of an icepack. Radiant, sensible and latent heat warms up the ice pack, the ice matrix melts, and with this process, water (with the isotopic loading) and the chemical substances will be washed out from the upper parts of the icepack, to be re-deposited where the water finally refreezes. Regarding water isotopes, the view is that there is no fractionation during melting of ice and that each crystal is homogenous in isotopic ratio, but water will be fractionated during freezing (Souchez and Lorrain, 1991), which then tends to alter the isotopic ratio of the water during transport. This leads to a change in isotopic load in the ice matrix where this water finally freezes. The icefield zone terminology by Benson (Paterson, 1994) is based on the amount of liquid water in the snow. The dry snow zone is the area where the snow never melts. The wet snow zone is the area where the melting is so effective that almost all of the pore space is filled with water at the end of the ablation season. The percolation zone is the area in between, where free water exists but never completely wets the snow pack.

Chemical signals are more sensitive to water transport than stable-water isotopic signals. This is because the chemical compounds, or charged ions, are usually situated in the perimeter of an ice crystal, within the grain boundaries between crystals. The reason for this is that the growing ice crystal refutes non-homogeneities when building its crystal lattice. This is the reason why there are microscopic water veins between three-grain intersections between ice crystals, since most salts are concentrated into this area. When melting an icepack, the crystal boundaries are the first areas where ice will melt, and widen up the grain boundaries, which washes out the chemical loading in the icepack from the level of deposition. When water refreezes during transport to colder zones on the icepack, all other non-water molecules will tend to be concentrated in the rest of the water. This means that the final water that freezes will have a very high integrated concentration of the chemical composition of the icepack through which the water has traveled. The final concentration of different compounds and charged ions is different between each species due to the solubility in water and their potential for incorporation in the ice lattice.

Ice structures represent a control we can use to investigate how much percolating water has re-worked the stratigraphic record. This is possible because the bubble volume in ice tells us if the ice has been infiltrated by water or not. An ice matrix that is created by firnification processes only (loss of pore space due to compaction and vapor-driven crystal growth) will have a large proportion of air bubbles within the ice crystals, while an ice matrix that has been suffering water infiltration, would tend to have a smaller amount of air bubbles. This is due to the simple reason that the infiltrating water will occupy the pores. This means that the amount of air bubbles in the ice can be used as a melt index. The more air bubbles as a percentage of all ice, the less melting has been affecting the ice, and the colder the ice has stayed (Koerner, 1977). Koerner uses the melt index as a measure of past summer temperatures. This may be valid for certain areas, but work by Pfeffer and Humphrey (1998) indicates that vapor transport, and creation of depth hoar crystals (that gives porous layers within the icepack) is probably a more important process.

Altogether this reveals the thorny path that atmospheric signals must travel, until the glaciologist retrieves a core from which climatologists can try to unravel its history. Most of the depositional and post-depositional problems mentioned here decrease with increased accumulation rates. This means that if the precipitation over the icefield is high enough, the distortion of the signal will decrease in weight. But, as areas with high precipitation are normally more maritime, and therefore warmer, the melt problem will increase. In the next section some examples are given to diagnose how severely this particular problem affects an ice core from the percolation zone of an Arctic icefield.

4. POTENTIAL TO RETRIEVE CLIMATIC INFORMATION FROM A PERIODICALLY-WET ICEFIELD: EXAMPLES FROM LOMONOSOVFONNA, SVALBARD

In the spring of 1997 a 121.6 m deep ice core was retrieved at the summit of Lomonosovfonna, situated in central Spitsbergen, within the Svalbard Archipelago (1255 m asl; 78°52' N, 17°26' E; Isaksson *et al.*, in press; Pohjola *et al.*, in press). The general climate of the area is dominated by dry arctic air, intersected by periods of maritime polar air advected from the southwest, with heat conveyed by the North Atlantic drift. The annual average near-surface temperature for Lomonosovfonna was estimated to be -12.5 °C by Soviet expeditions at 1020 m asl. (ftp.ngdc.noaa.gov/paleo/icecore/polar/svalbard/lomonosovfonna). The average annual temperature at the summit is probably about one degree lower.

The core was logged by DEP profiling, the ice structural information was optically read and described, and 5 cm samples were carefully prepared for analyses of stable isotopic content of the water and the chemistry of the samples (major ions; Jauhiainen *et al.*, 1999). The $\delta^{18}O$ content and the distribution of sodium and chloride are shown in Figure 1 as an example of the signal quality. The finding of a radioactive horizon at 18.7 m was attributed to the 1963 bomb horizon, and gave an average accumulation rate of 0.4 m a^{-1} at the drill site (Pinglot *et al.*, 1999).

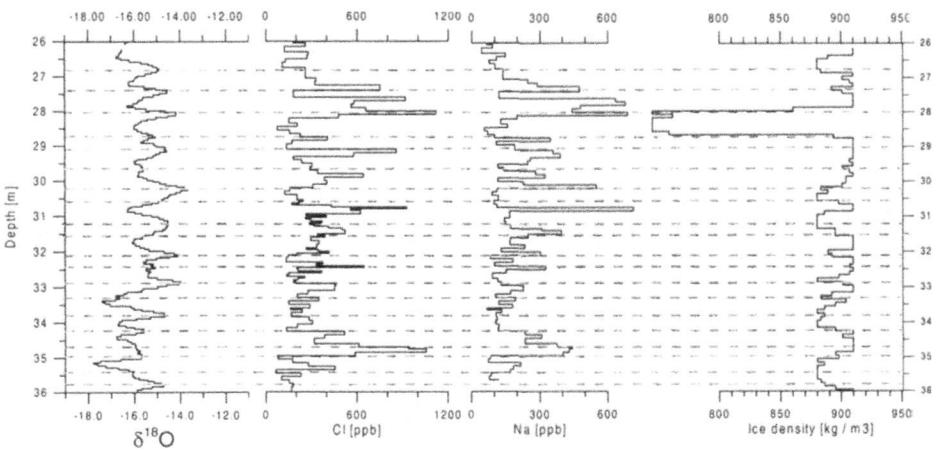

Figure 1. The distribution of $\delta^{18}O$, chloride ions, sodium ions and ice density from the depth interval, 26-36 m in the Lomonosovfonna Summit ice core. $\delta^{18}O$ and ion samples have a 5 cm resolution, the ice density data have a resolution of 10 cm. The dashed horizontal lines are depth levels interpreted as a summer signal. The interval is assumed to comprise 20 years.

Loggings of the borehole temperature in the summit core showed an average temperature of -2.8 °C and the 10 m temperature is ca. -2.3 °C (van de Wal, pers. comm.). The 10 °C difference between average air temperature and average ice temperature is mainly due to energy supply from two sources: conductive heat from geothermal sources and advective latent heat from percolating water, where geothermal heat flux can be neglected at the surface part of the ice pack. Clearly this site has been affected by melt. One way to investigate to what degree the ice pack has been influenced by processes driven by percolating water, is to examine the melt index for the site. Figure 2 shows the density of the Summit core and the melt index averaged over 5 m parts of the core. We find that the melt index varies with depth (time) and the total average is 41%. This is a lower value than that found in most other Arctic ice cores outside Greenland (Koerner, 1997). Notable is that the upper quarter of the core (approximately the post AD 1920 ice) shows a larger melt index. From the reading of ice structures along the core face we have a set of information that is dependent on climate, either proxy for summer melt or, following Pfeffer and Humphrey's (1998) suggestion, a proxy for temperature gradients in the snow pack (in fact a proxy for the amount of warm storms hitting the icefield).

This structural information can also be used to analyze the problems of post-depositional effects on the strata introduced by percolating water, by comparing the concentrations of the chemical species and isotopic fractions between ice layers created by refreezing water or by firnification processes.

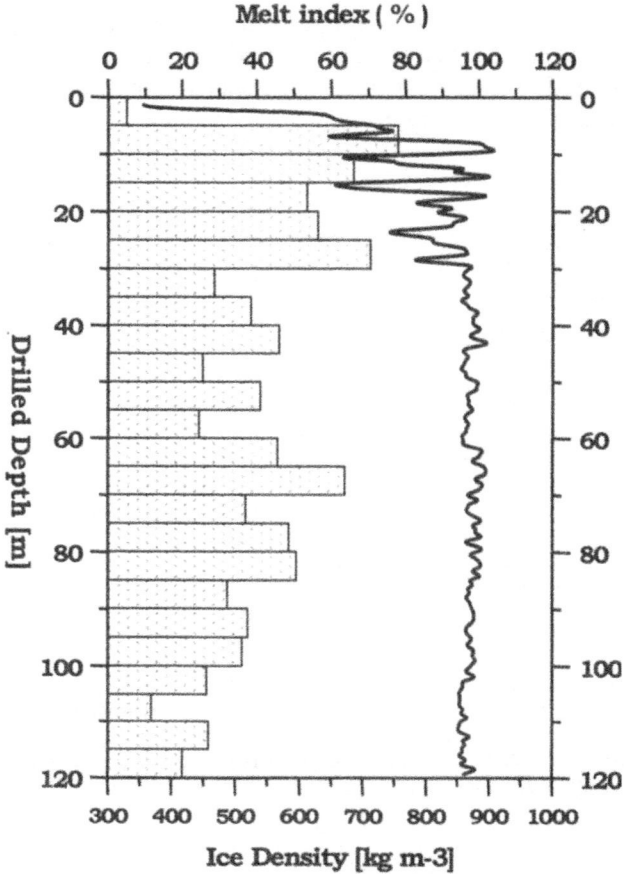

Figure 2. Distribution of the melt index (bars) and ice density of the Lomonosovfonna Summit ice core. The melt index is expressed in terms of relative mass.

Table 1 shows the relation between ice structures and some selected parameters in the depth interval 0-36 m. The distribution of ice structures with some of the parameters can be seen in the example in Figure 1. The ice structures are those parts in the density diagram with a density of 910 kg m^{-3}. The highest association found with ice structures in the chemical record was with the anions SO_4^{2-} and NO_3^- and the least affected were the cations NH_4^+ and K^+. This makes chemical sense, since the acids are more readily dissolved into the meltwater, and probably also more reluctant to be incorporated into the ice crystal lattice. The $\delta^{18}O$ from the upper quarter of the core showed no significant difference between the ice layer and the firn facies. There were ca. 0.5‰ more depleted isotopes in the bubble-free ice compared to the firn. The ice layers that are partly composed of infiltrating water that loses its heavy isotopes by refreezing of the water pool during the percolation path, show that the averaged effect of percolation is small. The difference in averaged isotopic composition between firn and ice layers is only 5 times the instrumental accuracy of detecting $\delta^{18}O$ values of the samples.

Table 1. Distribution of normalized concentrations of ions with respect to the different ice facies. Bold numbers are the average value, the standard deviation is given within parenthesis, and the number of samples is in italics. Each species is normalized to its average concentration. The ionic parameters are: chloride (Cl^-), sulfate (SO_4^{2-}), nitrate (NO_3^-), ammonium (NH_4^+), potassium (K^+), calcium (Ca^{2+}), and sodium (Na^+) The distribution of $\delta^{18}O$ given in the lower part of the table is not normalized, and shows the average value within each group in ‰ SMOW (Standard Mean Ocean Water). A third group (bubbly ice) in this study is not shown here.

Ion species	Normalized distribution of ions in each ice group	
	Firn	Bubble poor ice
Cl	**0.68** (0.46) *95*	**1.42** (0.92) *81*
SO4	**0.55** (0.47) *95*	**1.62** (1.08) *81*
NO3	**0.56** (0.42) *95*	**1.63** (1.03) *81*
NH4	**0.90** (1.64) *122*	**1.13** (0.90) *141*
K	**0.80** (1.66) *122*	**1.19** (1.14) *141*
Ca	**0.57** (1.12) *122*	**1.21** (1.78) *141*
Na	**0.73** (0.78) *122*	**1.21** (0.84) *141*
$\delta^{18}O$	**-15.51**(1.23) *250*	**-15.98**(0.91) *269*

A second diagnostic that can be used to check the damage done by percolating water to the initial stratigraphy, is to calculate cycles of the chemical species and of the $\delta^{18}O$ record. If it is assumed that $\delta^{18}O$ and the chemical species have a cyclical or periodical distribution within a year, then each positive peak value denotes the start of a new annual cycle. The number of assumed annual cycles can then be compared to a known reference horizon. In this case the number of cycles will be compared to the 1963 horizon. To perform this exercise one has to be careful of the resolution of the data; due to uncertainty and resolution problems in the analytical methods, real cycles may be hidden, or created accidentally. Therefore one needs to introduce a search window in which the amplitude of the wave needs to be larger than the detection limit, or the uncertainty of the parameter given from the laboratory methods. Another condition for the search window is the number of samples needed to form a wavelength. In this analysis the conditions were set to require at least 5 samples in order to form an annual cycle (25 cm of the record). Table 2 shows that the number of cycles found in the different datasets matches the number of years in the ice column fairly well.

Table 2. Number of cycles in each population within the depth interval 0-18.7 m. The search window for each population was based on the precision to determine analytically the value (amplitude), and that each half cycle must contain > 2 samples (wavelength). The depth, 18.7 m, is known to correspond to 1963, which means that 36 annual cycles should be present within the depth interval.

Parameter	$\delta^{18}O$	Na	K	Ca	NH4
Number of cycles	36	30	30	33	38

A third diagnostic is to compare the signal found in the ice pack with records sampled from direct measurements. The chemical data can be compared with concentrations measured at aerosol sampling stations, and isotopic data can be compared to rainwater sampling stations. In this paper, the isotopic record of the upper six meters containing the period winter 1989/90 to winter 1996/97 of the Lomonosovfonna record, is compared with the isotopic record for the same period from the coastal station Ny Ålesund, ca. 100 km west of Lomonosovfonna summit (IAEA/WMO, 1998). Table 3 shows the statistics from the two sites.

Table 3. Statistic comparison between ice-core data from Lomonosovfonna summit and monthly averages from the GNIP station at Ny Ålesund. The Ny Ålesund data covers the period 1990-1997 $\delta^{18}O$ values of precipitation. The ice-core data is from the upper 6 m of the ice core, which, based on $\delta^{18}O$ cycles, covers the period 1990-1997. The average value of Ny Ålesund data was offset -4.8‰ to normalize the average to the ice-core data. All values are in $\delta^{18}O$ (‰ SMOW). The diffused data is derived using methods described by Bolzan and Pohjola (2000), and by Whillans and Grootes (1985).

	Ice core	Ny Ålesund (-4.8)	Ny Ålesund, diffused 5 yr	Ny Ålesund diffused, 10 yr
Average	-16.0	-16.0	-16.1	-16.1
Minima	-19.1	-31.4	-21.2	-20.9
Maxima	-12.0	-10.2	-13.2	-14.0
1 Quartile	-17.4	-17.3	-16.9	-16.7
4 Quartile	-14.8	-14.5	-15.1	-15.3

There is a clear similarity of the statistical distribution between the two populations, offset by 4.8‰ by the Ny Ålesund data, which is probably due to the lower altitude of the coastal site, and the difference in geographical distance from the sea. The upper and lower quartiles of the data are similar, indicating that little post-depositional altering has changed the $\delta^{18}O$ signal in the ice pack. The maximum and minimum values are however much more extreme in the Ny Ålesund data. The reason why the extreme values may be lost in the ice pack data is probably due to the vapor-driven diffusion in the ice pack that smoothes the $\delta^{18}O$ data. To study these effects by diffusion, the Ny Ålesund data have been diffused by 5 and 10 years, using a numerical model described by Bolzan and Pohjola (2000; Table 3). From this, it is seen that the extreme values in the Ny Ålesund data migrate toward the average value, and to extreme values closer to the ice pack data found in the Lomonosovfonna ice core. (The diffusion of the $\delta^{18}O$ value is the reason why the seasonal range in $\delta^{18}O$ shown in Figure 1 is smaller than that given by the Ny Ålesund data).

These three diagnostic methods, the statistical analysis of the relation of peak concentration to infiltration ice, the numerical analysis by cycle counting different parameters, and the analytical analysis between ice pack data with real time sampled data, show that even in a core where 50% of the ice pack shows signs of refreezing of infiltrating water, there is a high potential for retrieval of information about atmospheric signals. At least 50% of the ice matrix in a piece of infiltration ice would still consist of the originally deposited snow, assuming that just before the infiltrating water occupied the pores, the firn had a density of ca. 500 kg m^{-3}. A condition for this is that even if there is an appreciable amount of melt in an ice pack, the signal will be stored as long as the meltwater percolates vertically and refreezes within the ice pack. In addition, if the percolating water refreezes within the annual or seasonal accumulation, then the time signal, even if somewhat vertically shifted, is stored within its stratigraphic layer. This means that the temperature of the ice pack and the accumulation rate are important parameters here; the colder winters and the larger accumulation on the icefield, the higher is the potential to save an atmospheric signal in a stratigraphic sequence in an ice pack that suffers periodical melt.

Obviously summer temperatures and the number of positive-degree days, play a role in determining melt, but these effects can be balanced by low winter temperatures and high accumulation rates. The creation of ice layers by infiltrating meltwater can even serve as a conservation process. The ice lens will act as an aquilude, an impenetrable zone for percolating water, and in such a way hinder water penetration of lower parts of the ice sequence. The ice layers also act as barriers for vapor transport in the ice pack, lowering the rate of vapor-driven diffusion appreciably. However, if melt is so severe that the icefield moves from the percolation zone into the wet snow zone, i.e. all pores in the snow/firn are filled with water, further vertical movement is slowed. This produces evaporation and lateral flow of water that damages the ice pack record. In this case, the infiltrating water creates superimposed ice to the ice surface of the last infiltration period. These ice packs are structurally identified by little or no frequency of bubbly ice, and in such stratigraphic sequences one must be very careful before drawing any significant conclusions of atmospheric signals in high temporal resolution from cores taken from this kind of ice pack.

In summary, this section shows that the damage done to stratigraphic records in an icefield suffering melt can be analyzed using three simple diagnostics, which give a measure of the degree of preservation of the deposited atmospheric signal.

5. POTENTIAL TO RETRIEVE CLIMATIC INFORMATION FROM THE SOUTHERN PATAGONIA ICEFIELD

The Southern Patagonia Icefield is a 13,000 km^2 icefield on the crest of the Andes spanning latitudes 48-51 °S and is ca. 350 km long and 30-40 km wide, centered along 73° 30' W, with unknown volume. The altitude of the surface is at approximately sea level for large outlets and > 3000 m asl for ice-covered peaks. The average altitude for the ice plateau is ca. 2000-1500 m asl (Warren and Sugden, 1993). The icefield is in the wet temperate climate zone. Climate data from Punta Arenas, situated at sea level ca. 300 km south southeast of the southern margin of SPI, show an average annual precipitation of ca. 0.5 m water, from 1990 to 1997. The average annual temperature of Punta Arenas during that period was 5 °C, and the warmest monthly average was January, with 11 °C (Figure 3). The climate of the SPI is not well known, but calculations and estimates of precipitation and temperatures are given elsewhere in this volume (Carrasco et al., 2002; Chapter 4). Based on, Carrasco's work (Carrasco et al., 2002; Chapter 4) the precipitation at SPI may be expected to vary between 1-15 m w.e. per year. The annual averaged temperature would give the zero isotherm at ca. 1000 m asl, the -5 °C isotherm at ca.

2000 m asl, and the elevation over 3000 m asl, is expected to be colder than -10 °C. Due to orographic effects, the eastern side of the icefield will be drier and colder than the windward western side. Precipitation and temperature patterns will, of course, have a strong relation to orographic and topographic characteristics. Figure 3 shows the temperature and isotopic distribution as an annual average from Punta Arenas between 1990 and 1997 (IAEA/WMO, 1998).

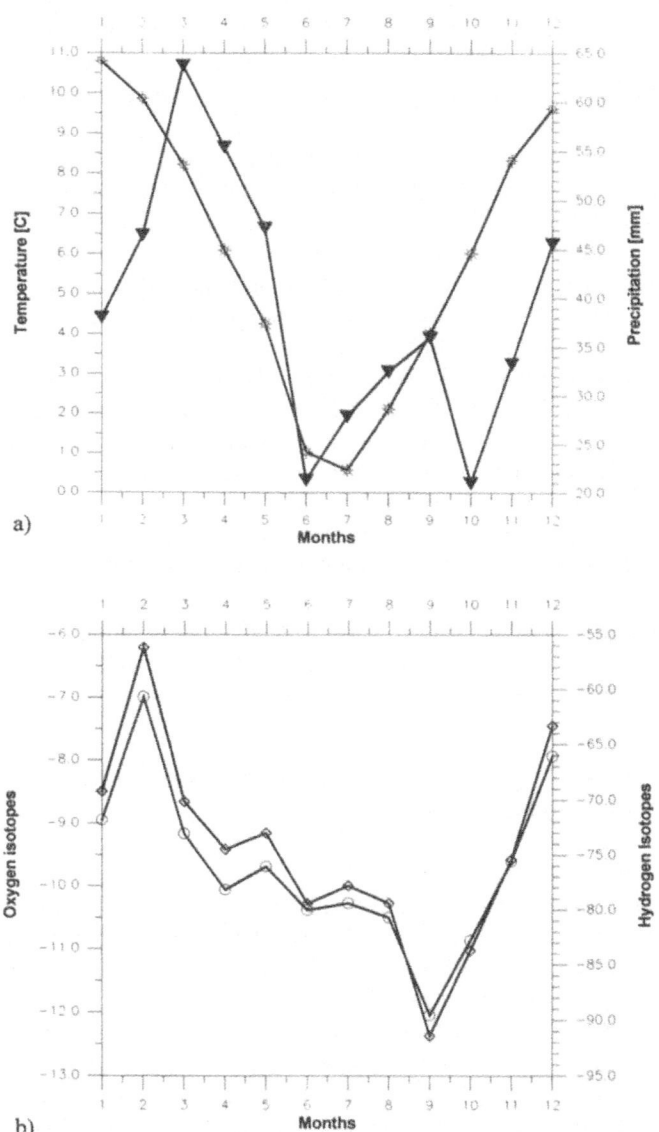

a)

b)

Figure 3. The monthly average distribution of a) air temperature (suns) and precipitation (triangles), and b) δ^{18}O (rings) and δD (diamonds) over Punta Arenas, during the period 1990-1997 (IAEA/WMO, 1998).

Ice cores have been taken on SPI by Aristarain and Delmas (1993), and by a Japanese-Chilean expedition in 1999 (Godoi *et al.*, 2002; Chapter 14). Ice cores have also been retrieved at the Northern Patagonia Icefield (HPN) by Matsouka and Naruse (1999) and by Yamada (1987). The cores from HPN were taken at 1500 and 1300 m asl, at ca. 47° S, 73.5° W. Both HPN sites were positioned at the wet snow zone, and were therefore heavily affected by percolated and refrozen water (the 1300 m asl core penetrated a thick aquifer). This would have washed out the isotopic signal from the firn pack. The chemical signals would have low atmospheric significance as well, since the wetting of the snowpack each year, and runoff, damages the signal. Other glaciological studies performed on the icefields are reviewed by Warren and Sugden (1993) and by Hardy and Bradley (1997).

In September 1986, Aristarain and Delmas (1993) retrieved a 13.2 m deep firn core at an altitude of 2680 m asl on the eastern side of the ice sheet (50°38' S, 73°15' W). Measurement of δD cycles and ice layers in the firn gave the conclusion that this core covered 5 years. This resulted in an average net accumulation rate of 1.2 m w.e. They also measured the major ions in the core. Their results show a broad range of ionic concentrations and that most of the 6 species have the same number of peaks as the δD data. An estimation of ice lenses from the structural illustration of the core indicates ca. 10-15% ice layers, giving a melt index of maximum 15%. Ice temperature in the bottom of their borehole was exactly 0 °C, while a calculation of the average air temperature showed -6 °C for this altitude. Clearly an appreciable amount of latent heat from refrozen water had been heating up the firn pack. This indicated that the core site was probably within the percolation zone, close to the boundary of the wet snow zone. The zero isotherm favors the wet zone, but preservation of the presumed atmospheric signals indicates that percolation has not been effective enough to wash out the signals. The relatively high accumulation at the site, probably prevents the whole firn pack getting wet during melt events, and preserves the annual signal within each years accumulation. This may be an explanation for the low melt index, despite the high temperature in the firn pack, as compared to the Svalbard ice core. Another explanation may be that some amount of the water may percolate down the strata to the firn/ice boundary, where it refreezes as superimposed ice.

In December 1999, a 46-meter deep core was recovered at the catchment of the Tyndall outlet glacier (50°59' S, 73°31' W; 1756 m asl). The average melt index for the core is ca. 10% (Godoi *et al.*, 2002; Chapter 14). The borehole temperature was approximately at the melting point and the finding of an aquifer at the firn/ice boundary at the bottom of the bore hole indicates that the firn pack was at 0 °C. This suggests that this site was in the wet snow zone. The core was drilled at the start of the summer, giving cold conditions in the upper part of the site, but warmer conditions in the lower part of the borehole. The water at the firn/ice boundary is typical for the wet snow zone, where water refreezes on the ice surface as superimposed ice. If this is the case, then some of the annually-deposited signal is transported downwards by some number of years. Results from isotopic and chemical analyses of the core will tell how affected the site is by this supposed percolation problem.

How likely is it that we might be able to recover good climate proxy records from SPI? In the case of water isotopes, Figure 3 shows that the area represented by Punta Arenas is influenced by an annual isotopic cycle, with seasonal maxima and minima in February and September respectively. The annually-averaged range in the δD data from

Punta Arenas is ca. 37‰. The average annual range in the core analysis by Aristarain and Delmas is ca 30‰ and similar to the Punta Arenas record. This indicates that Aristarain and Delmas' core record is not badly smoothed by percolation and diffusion processes at this depth of the firn pack. Both these smoothing processes are restricted by the relatively high accumulation rate. The chemical record from Aristarain and Delmas' core indicates that atmospheric signals are preserved. The association between ice layers and peak concentrations is not known, but due to the relatively high accumulation rate it is likely that most of the ions that leak out with percolating water will refreeze, and stay within the annual sequence.

6. DISCUSSION

What can be assessed from these ice-core records about the potential to retrieve climatic information from an SPI ice core? From the Svalbard ice core, it was found that atmospheric signals are preserved (even if they may be modulated by post-depositional processes) if the core site is high enough above the wet snow zone. To avoid the wetted snow zone one needs to consider an altitude > 2500 m asl, perhaps as high as 3000 m asl, in the southern part of SPI. Forthcoming results from the Tyndall core may tell if even lower altitude sites hold annual and seasonal signals, which is possible because of the large accumulation rates (Godoi *et al.*, 2002; Chapter 14). This limit is increasing to the western and northern part of the ice sheet, with the temperature distribution. Also, one needs to take the accumulation rate into consideration, where the driest parts may have too low an accumulation rate to be able to hold the annual signal within the annual accumulation. However, the results by Aristarain and Delmas indicate that an accumulation rate of 1.2 w.e. yr^{-1} may be enough to keep the water in its annual stratigraphical sequence at that altitude.

7. CONCLUSION

There is great potential for recovering good climate records from the Southern Patagonia Icefield, because high accumulation rates for the icefield favors preservation of signals within the strata. There will be problems with percolating meltwater at any given site, but selecting a site high enough (> 2500 m asl) will minimize the problem of wash-out and transposition of the deposited signal. Using the diagnostic scheme presented for the Svalbard core, will tell us how the records are affected by the percolation problem. All available climate data and the record of water isotope sampling by IAEA and chemical sampling of aerosols done in this region will, in this respect, be important information sources.

8. ACKNOWLEDGMENTS

I express my thanks to the Lomonosovfonna ice-core group for the hard labor they put into revealing the ice-core record, and for the discussion sessions on the scientific outcome. Special thanks to Elisabeth Isaksson, John Moore, Tonu Martma, and Rein Vaikmäe, for their work on the core data. Many thanks to CECS and the organizing committee for the generous arrangements for the workshop held on board the *Aquiles*. I also want to thank all the participants for the stimulating discussions at the plentiful dining 'n' science sessions during the workshop.

9. REFERENCES

Aristarain, A .J., and Delmas, R. J., 1993, Firn-core study from the Southern Patagonia ice cap, South America, *Journal of Glaciology*, **39**(132):249-254.

Bolzan, J. F., and Pohjola, V. A., 2000, Reconstruction of the undiffused seasonal oxygen isotope signal in central Greenland ice cores, *Journal of Geophysical Research*, **105**(C9):22095-22106.

Carrasco, J. F., Casassa, G., and Rivera, A., 2002, Meteorological and climatological aspects of the Southern Patagonia Icefield, in: *The Patagonian Icefields: a unique natural laboratory for environmental and climate change studies,* G. Casassa, F. V. Sepúlveda, and R. M. Sinclair, eds., Kluwer Academic/Plenum Publishers, New York, pp. 29-41.

Cuffey, K. M., and Stieg, E. J., 1998, Isotopic diffusion in polar firn: implications for interpretation of seasonal climate parameters in ice-core records, with emphasis on central Greenland, *Journal of Glaciology*, **44**(147):273-284.

Godoi, M. A., Shiraiwa, T., Kohshima, S., and Kubota, K., 2002, Firn-core drilling operation at Tyndall Glacier, in: *The Patagonian Icefields: a unique natural laboratory for environmental and climate change studies*, G. Casassa, F. V. Sepúlveda, and R. M. Sinclair, eds., Kluwer Academic/Plenum Publishers, New York, pp. 149-156.

Goodwin, I. D., Higham, M., Allison, I., and Jaiwen, B., 1994, Accumulation variation in eastern Kemp Land, Antarctica, *Annals of Glaciology*, **20**:202-206.

Hardy, D. R., and Bradley, R. S., 1997, Climate variability in the Americas from high elevation ice cores, Report of the IAI workshop held in Bariloche, Argentina, December 1996, University of Massachusetts, Amherst, U.S.A.

IAEA/WMO (1998). Global Network for Isotopes in Precipitation. The GNIP Database. Release 3, October 1999, URL: http://www.iaea.org/programs/ri/gnip/gnipmain.htm.

Isaksson, E., Pohjola, V., Jauhiainen, T., and 14 others, in press, A new ice record from Lomonosovfonna, Svalbard: viewing the data between 1920-1997 in relation to present climate and environmental conditions, *Journal of Glaciology*.

Jauhiainen, T., Moore, J., Perämäki, P., Derome, J., and Derome, K., 1999, Simple procedure for ion chromatographic determination of anions and cations at trace levels in ice-core samples, *Analytica Chimica Acta*, **389**:21-29.

Johnsen, S. J., Clausen, H. B., Cuffey, K. M., Hoffmann, G., Schwander, J., and Creyts, T., 2000, Diffusion of stable isotopes in polar firn and ice: the isotope effect in firn diffusion, in: *Physics of Ice Core Records*, T. Hondoh, ed., Hokkaido University Press, Sapporo, pp. 121-140.

Koerner, R M., 1977, Devon Island ice cap: core stratigraphy and paleoclimate, *Science*, **196**(1 April):15-18.

Koerner, R. M., 1997, Some comments on climatic reconstructions from ice cores drilled in areas of high melt, *Journal of Glaciology*, **43**(143):90-97.

Langway, C. C., 1967, Stratigraphic analysis of a deep ice core from Greenland, *US Army Cold Regions Research and Engineering Lab Research Reprints*, **77**:130.

Legrand, M., and Mayewski, P., 1997, Glaciochemistry of polar ice cores: a review, *Reviews in Geophysics*, **35**(3):219-243.

Matsuoka, K., and Naruse, R., 1999, Mass balance features derived from a firn core at Hielo Patagonico Norte, South America, *Arctic, Antarctic, and Alpine Research,* **31**(4):333-340.

Noone, D., Turner, J., and Mulvaney, R., 1999, Atmospheric signals and characteristics of accumulation in Dronning Maud Land, Antarctica, *Journal of Geophysical Research*, **104**(D16):19191-19211.

Nutterli, M. A., Röthlisberger, R., and Bales, R. C., 1999, Atmosphere-to-snow-to-firn transfer studies of HCHO at Summit, Greenland, *Geophysical Research Letters*, **26**(12):1691-1694.

Pasteur, E. C., and Mulvaney, R., 1999, Laboratory study of the migration of methane sulphonate in firn, *Journal of Glaciology*, **45**(150):214-218.

Paterson, W. S. B., 1994, *The Physics of Glaciers*, 3rd edition, Pergamon Press, Oxford.

Pfeffer, W. T., and Humphrey, N. F., 1998, Formation of ice layers by infiltration and refreezing of meltwater, *Annals of Glaciology*, **26**:83-91.

Pinglot, J. F., Pourchet, M., Lefauconnier, B., Hagen, J. O., Isaksson, E., Vaikmäe, R., and Kamiyama, K., 1999, Accumulation in Svalbard Glaciers deduced from ice cores with nuclear tests and Chernobyl reference layers, *Polar Research*, **18**(2):315-321.

Pohjola, V. A., Moore, J. C., Isaksson, E., Jauhiainen, T., van de Wal, R. S. W., Martma, T., Meijer, H. A. J., and Vaikmäe, in press, Effect of periodic melting on geochemical and isotopic signals in an ice core from Lomonosovfonna, Svalbard, *Journal of Geophysical Research*.

Souchez, R. A., and Lorrain, R. D., 1991, *Ice Composition and Glacier Dynamics*, Springer-Verlag, Berlin.

Stenberg, M., 2000, Spatial variability and temporal changes in snow chemistry, Dronning Maud Land, Antarctica, Ph. D. dissertation, Dept. of Physical Geography, Stockholm University, Sweden.

Warren, C. R., and Sugden, D. E., 1993, The Patagonian icefields: a glaciological review, *Arctic and Alpine Research*, **25**:316-331.

Werner, M., Mikolajewicz, U., Heimann, M., and Hoffmann, G., 2000, Borehole versus isotope temperatures on Greenland: seasonality does matter, *Geophysical Research Letters*, **27**(5):723-726.

Whillans, I. M., and Grootes, P. M., 1985, Isotopic diffusion in cold snow and firn, *Journal of Geophysical Research*, **90**(D2):3910-3918.

Yamada, T., 1987, Glaciological characteristics revealed by 37.6-m deep core drilled at the accumulation area of San Rafael Glacier, the Northern Patagonia Icefield, *Bulletin of Glacier Research*, **4**:59-67.

ARGENTINE PROGRAM ON CLIMATIC AND ENVIRONMENTAL STUDIES BY MEANS OF ICE CORES

Alberto J. Aristarain[1][*]

1. ABSTRACT

More than twenty years ago the Instituto Antártico Argentino initiated ice-core studies in the Antarctic Peninsula. It also carried out the first work of this type in the Southern Patagonia Icefield and preliminary investigations in the Argentine central Andes. Until recently, sample analyses were performed in two prestigious French laboratories in the frame of a joint cooperation. However, installation of the Laboratorio de Estratigrafía Glaciar y Geoquímica del Agua y de la Nieve, LEGAN (Laboratory of Glacier Stratigraphy and Water and Snow Geochemistry), in Mendoza, Argentina, has been recently finished and is equipped with the necessary facilities to develop glaciochemical determinations. Among our principal scientific results we mention the first isotopic-climatic reconstitution of the Antarctic Peninsula for the last 400 years. To investigate several millennia of high-resolution climatic and atmospheric fluctuations in this region we foresee completion, during the 2000/2001 austral summer, of an ice-core drilling to a depth of 350 m, initiated two years ago.

2. INTRODUCTION

The principal objective of the Icefields Meeting (Chile, March 24-29, 2000) was the preparation of a scientific program for the Southern Patagonia Icefield, including climatic recordings in the ice over the last few kilo years, which could be developed in the frame of an international cooperation led by Chile and Argentina.

With this purpose in mind, the present article aims to show briefly the ice-core activities in Argentina, which are developed by the LEGAN of the Instituto Antártico

[1] Laboratorio de Estratigrafía Glaciar y Geoquímica del Agua y de la Nieve (LEGAN), Instituto Antártico Argentino, Casilla 131, 5500 Mendoza, Argentina.

[*] corresponding author: aristar@lab.cricyt.edu.ar

The Patagonian Icefields: A Unique Natural Laboratory for Environmental and Climate Change Studies.
Edited by Gino Casassa et al., Kluwer Academic /Plenum Publishers, 2002.

Argentino (IAA). In consequence, we present some results previously published and we indicate the corresponding references. Finally, we report some of our current studies.

2.1. Natural Archives

It is becoming increasingly important to know the climate mechanisms in order to forecast their future evolution. Then, it is necessary to separate the natural variability from that induced by human activities, and this implies the interpretation of environmental information prior to instrumental series. To this end, the high latitude icefields and high mountain glaciers are privileged archives. On the one hand, stable isotopes in the snow precipitation (δD, $\delta^{18}O$ and excess deuterium) relate principally to air temperature. On the other hand, the snow includes particles, gases and aerosols originating from natural processes (volcanic sulfate, sea salt, organic compounds, etc.) or from human activity (CO_2, nitrates, heavy metals, exhaust gases, etc.). In cold sites, where melting snow is negligible or absent, these environmental proxies are deposited chronologically without mixture. They permit the reconstruction, year by year, of the climate and the atmospheric composition, on a regional or global scale, for thousands of years. It is also possible to investigate their relationship to, for example, global warming and the increasing emission of greenhouse gases.

2.2. Ice-Core Studies in Argentina

Ice-core studies in Argentina were originated as a consequence of the international program Glaciology of the Antarctic Peninsula (GAP) created with the participation of glaciologists from the 5 nations interested in this region (SCAR, 1974). This program not only highlights the interest on the Antarctic Peninsula but also on the Patagonian Icefields for this type of study (Figure 1). As part of the agreement derived from GAP, the IAA obtained 10 m depth of firn cores in James Ross Island ice cap (Figure 1) during the 1974/1975 austral summer. Since then, ice- and firn-core studies in Argentina have been principally supported by the IAA and Consejo Nacional de Investigaciones Científicas y Técnicas (CONICET) and have been developed jointly with the French laboratories, the Laboratoire de Glaciologie et Géophysique de l'Environnement and the Laboratoire des Sciences du Climat et de l'Environnement, where we carried out chemical analyses and participated in isotopic determinations. LEGAN, in Mendoza, has just been completed with the financial support of the IAA and CONICET. It also received contributions from local and international organizations (European Union and Inter-American Institute grants). LEGAN is equipped with clean and cold rooms and chemical analytical equipment for determination of ionic concentrations of soluble impurities in the samples (Figure 2).

3. RESULTS

The Antarctic Peninsula and adjacent islands would make it possible to connect long ice-core series from central Antarctic with high-resolution records from Patagonian icefields. This region is of great interest due to its extreme climatic sensitivity, probably

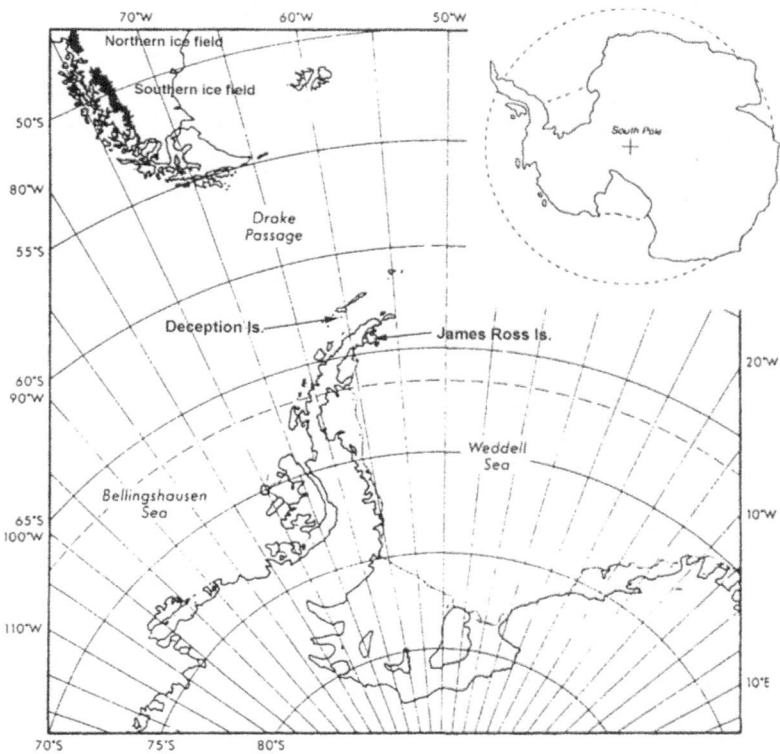

Figure 1. Map of the Antarctic Peninsula region and of southern South America. James Ross Island, Deception Island and the Patagonian icefields are indicated.

Figure 2. View of the ion chromatograph and the atomic absorption spectrophotometer, to measure ionic concentration of the samples, in the clean chamber of the LEGAN.

related to the extent of variable sea-ice cover. Furthermore, the existence of the long and relatively numerous meteorological series makes it possible to calibrate climate proxies determined in annually dated ice cores.

Among the principal results obtained on James Ross Island ice cap, where we have performed detailed investigations (64°12'54'' S and 57°40'30'' W; 1640 m asl), we mention the first isotope-climatic reconstitution for the Antarctic Peninsula, which covers the last 400 years with high resolution (Aristarain *et al.*, 1990; Figure 3).

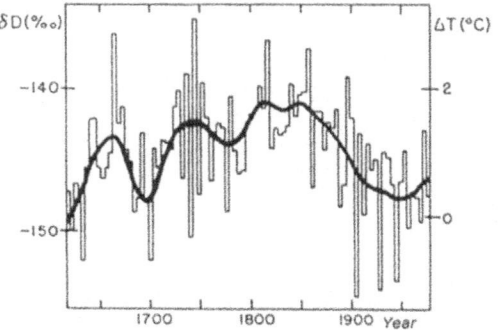

Figure 3. Isotope climatic record obtained in James Ross Island ice cap for the period 1607-1980. Aristarain, A. J. *et al.*, 1990, A 400 year isotope record of the Antarctic Peninsula climate, *Geophysical Research Letters*, **17**(12):2369-2372. Reproduced by permission of American Geophysical Union.

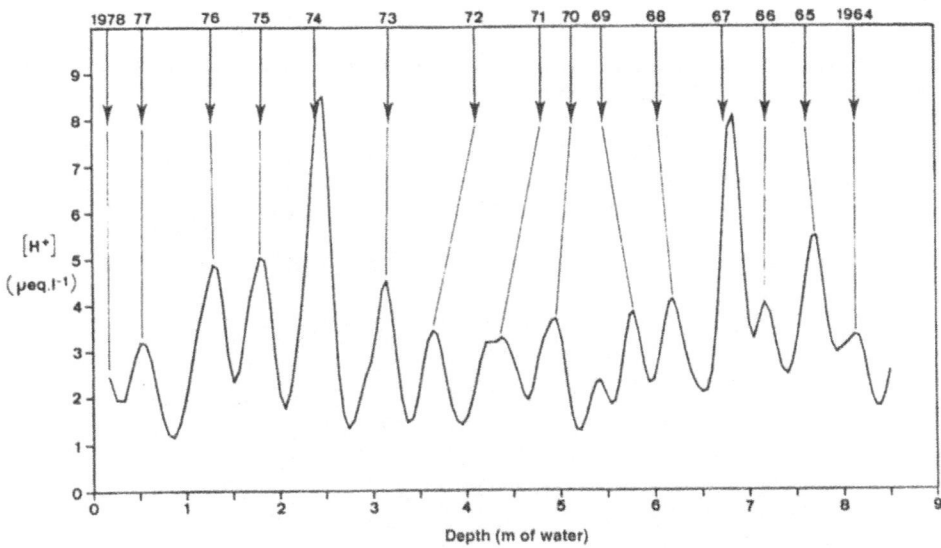

Figure 4. Annual oscillations of acidity (principally H_2SO_4) along a 16 m firn core of James Ross Island, Antarctica (about 8.5 m water equivalent). The peak in 1967 corresponds to a volcanic eruption in Deception Island, indicated in Figure 1.

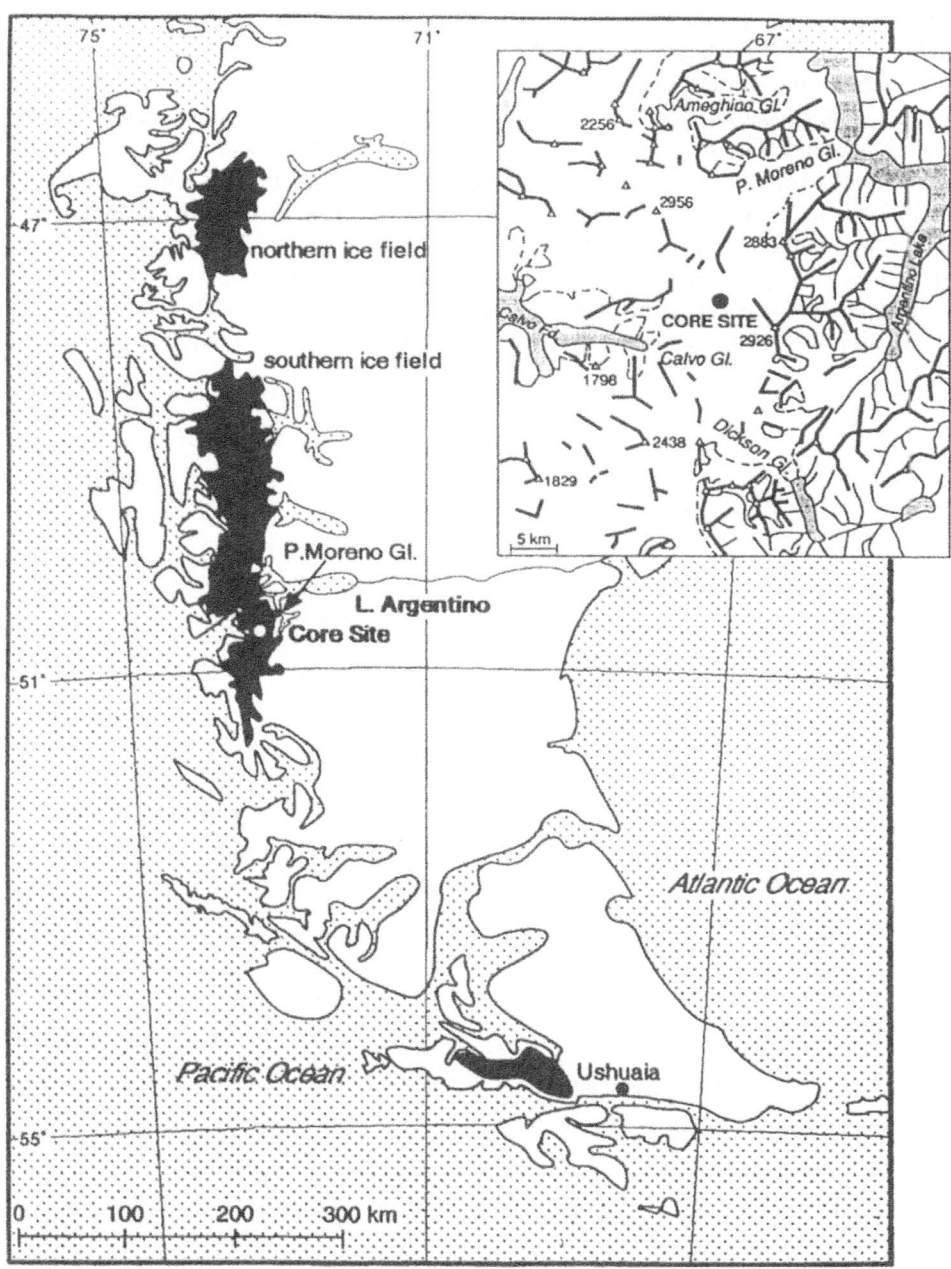

Figure 5. Map of southern South America including the two Patagonian icefields and the drilling site (core site).

The amplitude in °C of isotopic fluctuations has been estimated by means of a transfer function determined by comparing annual means of δD and temperatures from neighbor stations (Aristarain *et al.*, 1986). In relation to glaciochemical analyses, we could mention the marked seasonality we found in acidity and sulfate, which were used later for detailed dating, and the detection of the 1967 volcanic activity of the nearby Deception Island (Figures 1 and 4: Aristarain, 1980; Aristarain *et al.*, 1982). Later, we found an unknown eruption of this volcano dated 1642 (Aristarain and Delmas, 1998).

In 1986 we performed an ice-core drilling to a depth of 13 m in the accumulation zone of the Southern Patagonia Icefield (50°38' S and 73°15' W; 2680 m asl: Aristarain and Delmas, 1993), where several important glaciers originate (Figure 5).

Temperature measurements showed that the icefield is temperate. The chemical profiles indicate that some soluble impurities have been partly washed out by percolation (Figure 6).

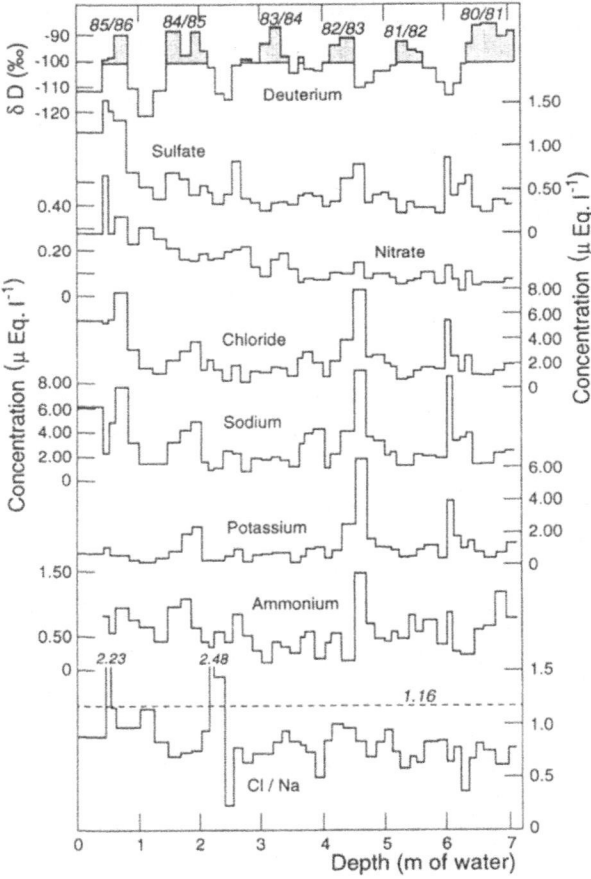

Figure 6. Ion concentration versus depth (in m water equivalent) of our Southern Patagonia Icefield core. The time-scale is given by the isotope profile. Reprinted from the Journal of Glaciology with permission of the International Glaciological Society.

On the contrary, deuterium content is well preserved down to the bottom of the core and we interpreted the profile as seasonal oscillations, which enabled us to calculate the annual accumulation of snow (Figure 7). Nevertheless, recent glaciological work carried out by glaciologists from Japan, Chile and Argentina near the ice divide on Tyndall Glacier, shows some disagreement with our estimations.

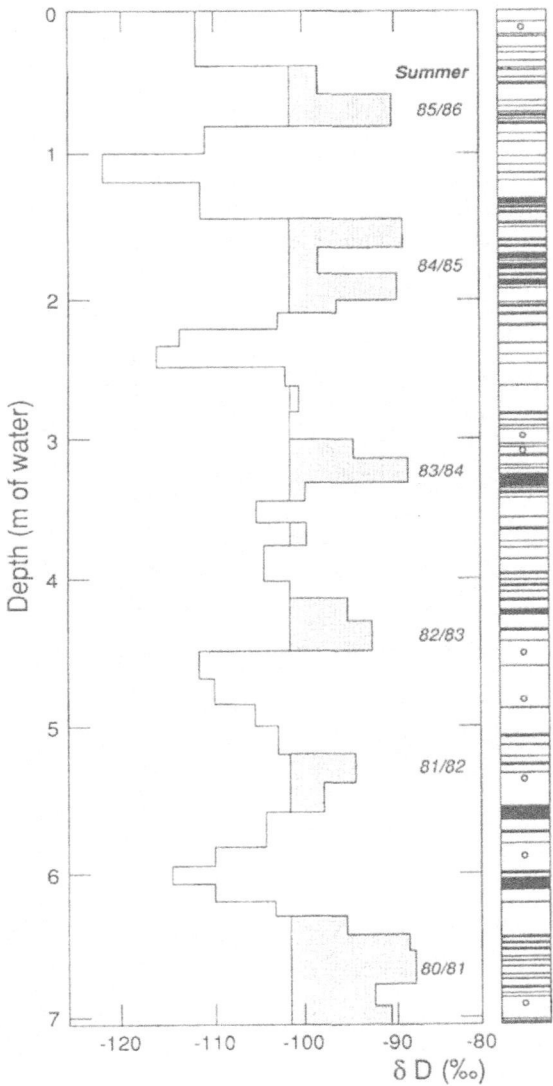

Figure 7. Variations of the deuterium content, which were interpreted as annual oscillations, compared with the snow stratigraphy (black segments represent ice layers) corresponding to the core site in the Southern Patagonia Icefield. Reprinted from the Journal of Glaciology with permission of the International Glaciological Society.

In the south-central Andes we selected potential sites for preliminary work to fill the current knowledge gap between ice-core series in South American tropical glaciers and the Patagonian Icefields. In 1997, fieldwork was carried out on one of these glaciers, Mesón San Juan (~33°33' S and 69°46' W; 6035 m asl), near the city of Mendoza. Nevertheless, it was not possible to drill due to the existence of a field of large penitentes and crevasses, a situation that had not been observed in previous summers. It is therefore important to start drilling on other possible glaciers in this area.

4. PRESENT WORK

We are currently investigating the relation between climate, the extent of sea ice cover and the chemical content of the snow in the northern area of the Antarctic Peninsula. We are also undertaking new measurements of stable isotopes and excess deuterium to validate the isotopic temporal series from this region as climate proxies.

Figure 8. Ice-core activities on James Ross Island ice cap, Antarctic Peninsula. The drilling operations were developed in a large trench dug in the snow, used as protection from the extreme climatic conditions.

During the 1997/1998 Argentine Antarctic Campaign, we carried out two drillings, to depths of 117 and 122 meters on James Ross Island ice cap (Figure 8) in the frame of a joint cooperation between Argentina, Brazil, Canada and France. This sampling is planned to be completed to the bedrock at a depth of 350 m, during the 2000/2001 austral summer. We expect to obtain several millennia of high-resolution climatic and atmospheric fluctuations, according to our chronological model (Aristarain, 1980). As in previous campaigns, the fieldwork will be directed by the Argentine component, which will also provide the logistic support.

5. CONCLUDING REMARKS

Ice-core studies would be very valuable for environmental reconstitution in the Southern Patagonia Icefield. However, the cost of a drilling operation to extract the cores is high, especially due to the necessary logistics for working in the extreme conditions of this region. In addition, sample analyses require sophisticated laboratories with clean and cold rooms. For these reasons, an eventual international joint program should make use of the potentialities of each scientific group interested in participating. It should also enable the development of glaciological research in South American laboratories, where the economic situation is a considerable handicap.

6. REFERENCES

Aristarain, A. J., 1980, Etude glaciologique de la calotte polaire de l'Ile James Ross, Péninsule Antarctique. Thèse de 3ème. cycle, *Publication No. 322 du Laboratoire de Glaciologie et Géophysique de l'Environnement du CNRS*, Grenoble, France (in French).

Aristarain, A. J., Delmas, R. J., and Briat, M., 1982, Snow chemistry on James Ross Island, (Antarctic Peninsula), *Journal of Geophysical Research*, **87**(C13):11004-11012.

Aristarain, A. J., Jouzel, J., and Pourchet, M., 1986, Past Antarctic Peninsula climate (1850-1980) deduced from an ice-core isotope record, *Climatic Change*, **8**(1):69-89.

Aristarain, A. J., Jouzel, J., and Lorius, C., 1990, A 400 year isotope record of the Antarctic Peninsula climate, *Geophysical Research Letters*, **17**(12):2369-2372.

Aristarain, A. J., and Delmas, R. J., 1993, Firn-core study from the Southern Patagonia ice cap, South America, *Journal of Glaciology*, **39**(132):249-254. .

Aristarain, A. J., and Delmas, R. J., 1998, Ice record of a large eruption of Deception Island volcano (Antarctica) in the XVIIth century, *Journal of Volcanology and Geothermal Research*, **80**:17-25.

SCAR, 1974, SCAR Bulletin. Report of the SCAR Executive Meeting, Cambridge, 9-11 July 1973. *Polar Record*, **17**(106):79-99.

FIRN-CORE DRILLING OPERATION AT TYNDALL GLACIER, SOUTHERN PATAGONIA ICEFIELD

María Angélica Godoi[1*], Takayuki Shiraiwa[2], Shiro Kohshima[3], and Keiji Kubota[2]

1. ABSTRACT

A 45.97 m-deep drilling operation was carried out during November/December 1999 at the accumulation area of Tyndall Glacier, 1756 m asl, in the southern end of the Southern Patagonia Icefield. In the field, the firn-core obtained was the subject of stratigraphic observations and density measurements. Preliminary results suggest an extremely high accumulation (about 13.5 m w.e.), which agrees with snow measurements observed during the fieldwork. Laboratory analyses are currently underway in Japan to characterize the physical, chemical and biological properties of the core.

2. INTRODUCTION

Ice cores are nowadays widely acknowledged as unique archives of past atmospheric and environmental conditions. Although a good number of cores have been retrieved from polar, tropical and subtropical areas, recording valuable information about various global and local processes, little has been done in this respect in middle latitudes, particularly in the Patagonian icefields.

Two shallow cores have been recovered from the Northern Patagonia Icefield (NPI, Figure 1), both of them for glaciological studies. Yamada (1987) reported the drilling of a firn/ice core to the depth of 37.6 m in the accumulation area of San Rafael Glacier (1296 m asl), in November 1985. An annual net accumulation of 3.45 meters expressed as water equivalent (m w.e.) was estimated for the year 1984. Another drilling operation (Matsuoka and Naruse, 1999) recovered a 14.5 m-deep firn-core, in November and December 1996, in the accumulation area of Nef Glacier (1500 m asl). They reported a net accumulation of 2.2 m w.e. in the calendar year 1996.

[1] Instituto de la Patagonia, Universidad de Magallanes, Casilla 113-D, Punta Arenas, Chile; [2] Institute of Low Temperature Science, Hokkaido University, Sapporo 060-0819 Japan; [3] Faculty of Bioscience and Biotechnology c/o Faculty of Science, Tokyo Institute of Technology, Tokyo 152, Japan.

* corresponding author: mangel@ona.fi.umag.cl

The Patagonian Icefields: A Unique Natural Laboratory for Environmental and Climate Change Studies.
Edited by Gino Casassa et al., Kluwer Academic /Plenum Publishers, 2002. **149**

The Southern Patagonia Icefield (SPI, Figure 1), located between latitudes 48° 20' S and 51° 30' S, is one of the largest ice masses in mid-latitudes, covering an approximate area of 13,000 km² (Aniya *et al.*, 1996). Despite this fact, little is known about its glaciology and climate. Previous to the present work, only one shallow firn-core had been recovered from this ice cap: a 13.17 m firn-core drilled on a plateau (2680 m asl) at the headwater of Perito Moreno Glacier in September 1986 (Aristarain and Delmas, 1993). Based on the visual stratigraphy and isotopic profile, they estimated a mean annual accumulation of 1.2 m w.e. for the period between the summer of 1980/81 and 1985/86.

This work reports the firn-core drilling to a depth of 45.97 m, carried out by a Japanese-Chilean joint expedition at the accumulation area of Tyndall Glacier in November/December 1999. The main objective of the project is to assess the recent mass balance of this glacier.

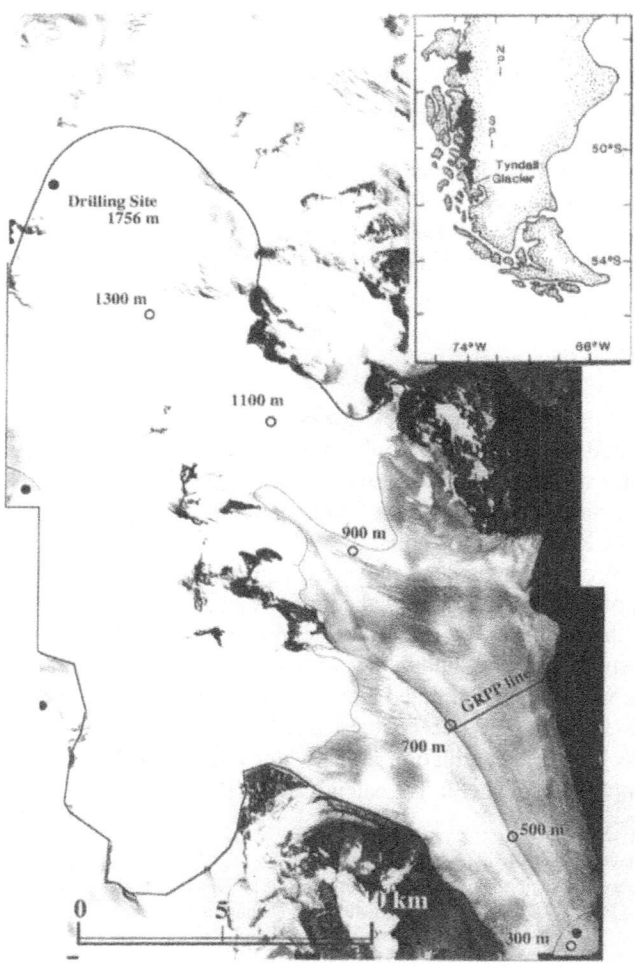

Figure 1. Location of the drilling site.

3. DRILLING SITE

The drilling site was located in the accumulation area of Tyndall Glacier, 50° 59' 05" S, 73° 31' 12" W, on a relatively flat plateau at 1756 m asl, near the ice divide (Figure 1). Tyndall Glacier flows southwards in the southernmost part of SPI, covering a total area of about 331 km^2, stretching for 32 km and calving into a proglacial lake (Aniya et al., 1996). Climate in Patagonian glaciers is dominated by the westerlies which transport moisture from the Pacific Ocean, depositing large amounts of rain and snow on the windward side of the Andean range, which runs north-south, perpendicular to the Southern Hemisphere westerlies in this area (Warren and Sugden, 1993; Takeuchi et al., 1999). There are no meteorological records from the accumulation areas of SPI. During the field campaign, no more than four days out of 25 presented good visibility. The camp was constantly threatened with blizzards and heavy snow, and clearing snow from the tents was a hard daily task. The level of the surface was continuously covered by fresh snow, suggesting an extremely high accumulation.

4. LOGISTICS

People and equipment were transported by helicopter from the headquarters of Torres del Paine National Park, 40 km east of the drilling site, on 29 November. Although the drilling operation was completed by 9 December, the expedition extended beyond the planned date since weather conditions did not allow the helicopter to perform the evacuation flight. On 23 December, with terrestrial support from the lower part of the glacier, people and minimum equipment were evacuated by snowmobile to a lower altitude (1100 m asl), from where conditions allowed for evacuation by helicopter. Most of the equipment carried to the plateau had to be abandoned, becoming buried under several meters of snow.

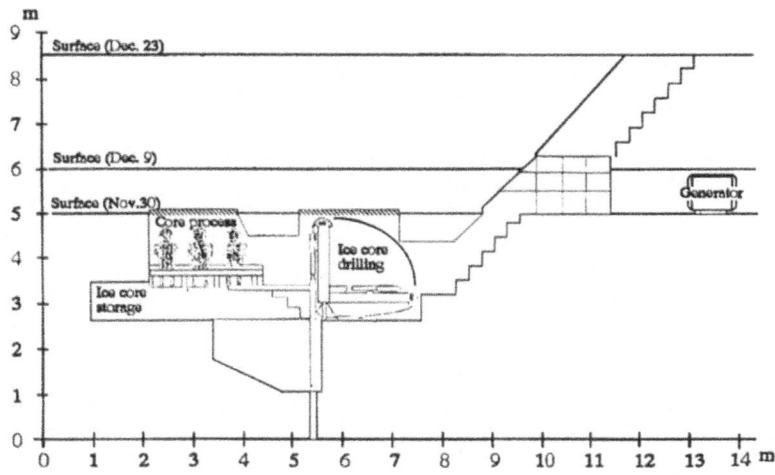

Figure 2. Cross profile of the drilling trench.

A trench was dug to allow the drilling operation to be carried out under any weather condition. Enough room was allowed for core drilling, processing, and storage of the samples (Figure 2). Excavation of the drilling trench and installation of the drilling system took about 5 days.

A firn-core was obtained up to a total drilling depth of 45.97 m. This comprised a 40.65 m core of the deeper layers, recovered in about 5 days, using an electromechanical drill, and a shallow core of the surface layers, 5.47 m long, obtained independently using a hand auger. Both cores superimpose partially by a length of 15 cm.

The lightweight waterproof electromechanical drill used (Figure 3), so-called "dokodemo-drill" ("drill for wherever you want"), was manufactured at the Institute of Low Temperature Science, Hokkaido University, Japan, and is especially designed for high-mountain temperate glaciers having a water aquifer at the firn/ice transition zone. The drilling system is driven by a 600 W, 200 V, variable speed AC motor (20-40 cm/s), and included a newly designed pantographic-type of anti-torque mechanism which presented some problems during the campaign. The mast consists of a tilting tower 2.3 m high and the variable speed winch is able to lift an 80 kg load, at an average speed of 15 m/min. The winch uses a 1 kW, 100 V, AC motor which is supplied by a variable voltage controller.

Figure 3. The "dokodemo-drill" in operation.

This drill is able to retrieve a core of 74 mm in diameter with a length of up to 80 cm for firn and 65 cm for ice. In the field, the length of the core obtained at each drilling run was, on average, 50 cm (Figure 4), except when problems arose with the pantographic mechanism.

The core recovered was cut into 25 cm sections. After stratigraphic observations and bulk density measurements the core was cut vertically into 3 pieces (one half, and two fourths). The half was packed in sealed plastic bags and kept untouched. The external part of the two other pieces was removed using pre-cleaned tools and the samples were packed in separate sealed plastic bags. Unfortunately, as weather conditions only allowed for an emergency evacuation procedure, only one set of samples (one fourth of the core) could be saved for laboratory analyses. Those samples were kept frozen until their arrival in Japan, where they are currently being analyzed. We had planned to measure the borehole temperature after the drilling, but this was not possible due to the presence of meltwater inside the hole.

5. CORE STRATIGRAPHY

The core recovered consists of dry firn with ice layers of various thickness (1-50 mm) until water soaked firn started to be obtained at a depth of 42.55 m. Distribution of the ice layers in the core was not uniform. Some sections of the core (4-5 m, 18-24 m, and 38-39 m in depth) contained many thick ice layers, while others contained very few ice layers.

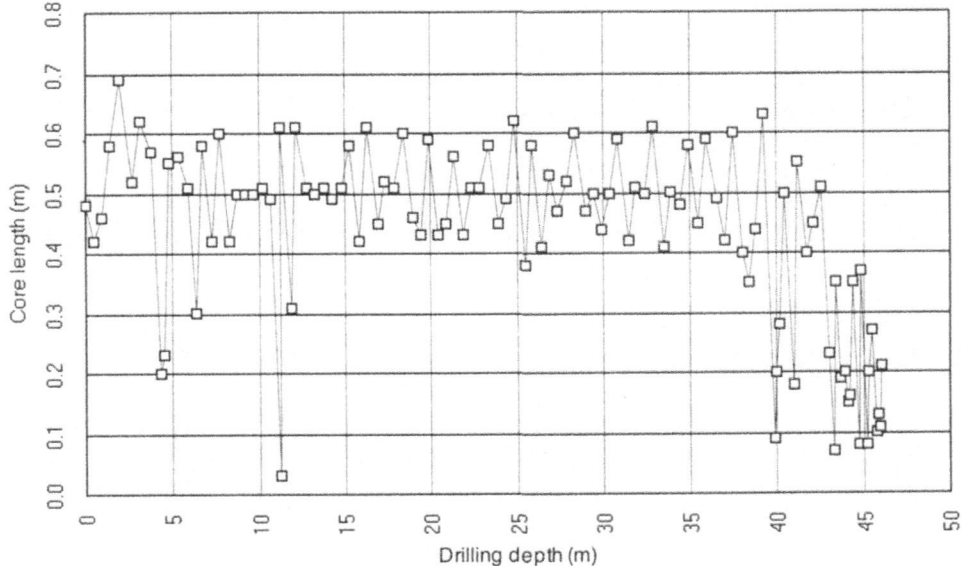

Figure 4. Length of the core at each drilling run.

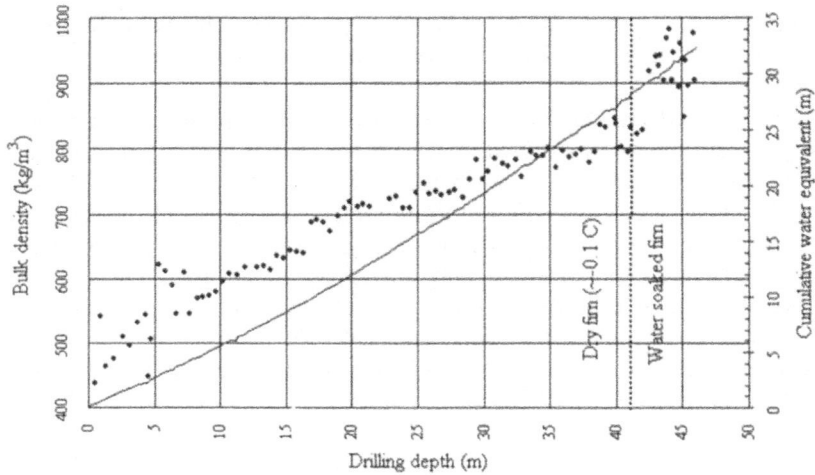

Figure 5. Profiles of bulk density and cumulative water equivalent (solid line) in the firn-core.

6. BULK DENSITY

Bulk density was calculated from the weight and volume of every core section obtained. As can be seen in the profile (Figure 5), density presents a nearly linear and slow increase with depth, until about 42 m when water-soaked firn started to be obtained. Then it rises quickly reaching values close to the density of ice. Pure ice was still not reached at a depth of 45.97 m where the drilling operation stopped.

Thus, the firn/ice transition zone for this glacier started at a depth of 42 m, and the ice boundary can be estimated according to the trend of the density data (Figure 5) at around 50 m in depth. Yamada (1987) found this transition zone between the depths of 19.7 and 26.7 m at the accumulation area of San Rafael Glacier (NPI), 1296 m asl. At Tyndall Glacier this transition zone started much deeper probably due to a higher accumulation rate. The solid line in Figure 5 represents the cumulative water equivalent profile.

7. MELT FEATURES

The number and thickness of ice layers found in a given piece of core is proportional to the degree of melting during the corresponding period of time and can be considered an indication of summer warmth. Melt feature percentage is an index indicating the fraction between ice and firn along the core, and it has been used as a summer-temperature proxy record (Koerner, 1977; Paterson, 1994).

Melt feature percentage (Figure 6) was calculated in terms of water equivalent for every 1m of core. Several ice layers of variable thickness were found along the firn-core. Their interpretation as melt-freeze events occurring at the surface is not straightforward for temperate glaciers.

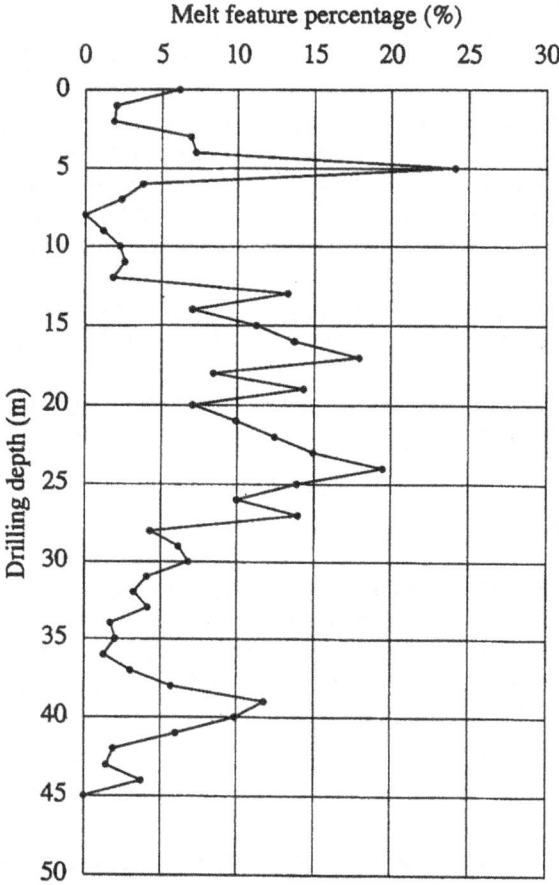

Figure 6. Melt feature percentage in the 45.97 m firn-core. The first meters of snow accumulated during the field work, as seen in Figure 2. Thus, the melt peak at 5 m corresponds to the present summer at the time of the drilling. The 0 m level corresponds to December of 1999, assumed to be mid-summer of 1999/2000. Winters are assumed to correspond to low levels of melting. Therefore, in between, the levels of 20 m and 39 m are thought to correspond to mid-summer of 1998/99 and 1997/98, respectively. This interpretation has been confirmed by subsequent isotopic analysis (Shiraiwa *et al.*, in press).

Ice layers might be the result of percolation of liquid precipitation as well as percolation of surface melting during summer or wintertime, finally re-freezing at depth within a layer of a previous season. However, assuming that those effects are, in this case, not enough to mask the signal, we may consider that the middle of the summers 1999/2000, 1998/99, and 1997/98 correspond to the depth of 0 m, 20 m, and 39 m respectively. Thus, using the graph of cumulative water equivalent (Figure 5), we can obtain a preliminary estimation of 13.5 m w.e., for the annual accumulation during these years. It is expected that in this case, meltwater percolation does not destroy the seasonal patterns in the isotopic profiles, so that this figure can be checked when the core analyses are completed.

8. CONCLUSIONS

The drilling in the accumulation area of Tyndall Glacier (1756 m asl) showed temperate firn down to a depth of 45.97 m. Water-saturated firn was found at approximately 42 m, marking the beginning of the firn/ice transition zone. Thus, the firn/ice boundary is below 46 m, and is expected to be around a depth of 50 m.

Preliminary results suggest an annual accumulation of about 13.5 m w.e., for the years between the summers 1999/00, 1998/99, and 1997/98. This is an extremely high value when compared to the results of Aristarain and Delmas (1993), of 1.2 m w.e., Yamada (1987), with 3.45 m w.e. and Matsuoka and Naruse (1999), with 2.20 m w.e. However, it is in good agreement with the conditions of snow precipitation experienced during the field campaign. This figure has to be checked with subsequent core analyses currently underway.

Weather conditions experienced in the field proved that logistics are critical when planning a drilling operation in SPI.

9. ACKNOWLEDGMENTS

The authors would like to thank Marcelo Arévalo and Jorge Quinteros for their valuable support during the field campaign. We also thank Gino Casassa and Andrés Rivera for leading the emergency evacuation of the team. This research was funded by the Ministry of Education and Culture of Japan, and the Fondo Nacional de Ciencia y Tecnología (FONDECYT 1980293) of Chile.

10. REFERENCES

Aniya, M., Sato, H., Naruse, R., Skvarca, P., and Casassa, G., 1996, The use of satellite and airborne imagery to inventory outlet glaciers of the Southern Patagonia Icefield, South America, *Photogrammetric Engineering and Remote Sensing*, **62**(12):1361-1369.

Aristarain, A. J., and Delmas, R. J., 1993, Firn-core study from the Southern Patagonia ice cap, South America, *Journal of Glaciology*, **39**(132):249-254.

Koerner, R. M., 1977, Devon Island ice cap: core stratigraphy and paleoclimate, *Science*, **196**(4285):15-18.

Matsuoka, K., and Naruse, R., 1999, Mass balance features derived from a firn-core at Hielo Patagónico Norte, South America, *Arctic, Antarctic and Alpine Research*, **31**(4):333-340.

Paterson, W. S. B., 1994, *The Physics of Glaciers*, Third Edition, Pergamon, Oxford, pp. 393-394.

Shiraiwa, T., Kohshima, S., Uemura, R., Yoshida, N., Matoba, S., Uetake, J., and Godoi, M. A., in press, High net accumulation rates at the Southern Patagonia Icefield revealed by analyses of a 45.97-m-long ice core, *Annals of Glaciology*, **35**.

Takeuchi, Y., Naruse, R., Satow, K., and Ishikawa, N., 1999, Comparison of heat balance characteristics at five glaciers in the Southern Hemisphere, *Global and Planetary Change*, **22**:201-208.

Warren, C. R., and Sugden, D., E., 1993, The Patagonian Icefields: a glaciological review, *Arctic and Alpine Research*, **25**(4):316-331.

Yamada, T., 1987, Glaciological characteristics revealed by 37.6-m deep core drilled at the accumulation area of San Rafael Glacier, the Northern Patagonia Icefield, *Bulletin of Glacier Research*, **4**:59-67.

FIRST RESULTS OF A PALEOATMOSPHERIC CHEMISTRY AND CLIMATE STUDY OF CERRO TAPADO GLACIER, CHILE

Patrick Ginot[1,2], Margit Schwikowski[1*], Heinz W. Gäggeler[1,2], Ulrich Schotterer[1,3], Christoph Kull[4], Martin Funk[5], Andrés Rivera[6,7], Felix Stampfli[8], and Willi Stichler[9]

1. ABSTRACT

In February 1999 a 36 m ice core reaching bedrock of the cerro Tapado summit glacier (5550 m, 30°08' S, 69°55' W) was recovered in order to investigate the suitability of this glacier as paleoenvironmental and climate archive. Site selection was based on the assumption that this area is strongly influenced by the El Niño phenomenon. Glaciochemical data indicate that a record of about 100 years is contained in the ice core and that El Niño periods are characterized by low concentrations of chemical species.

2. INTRODUCTION

The central Andes are a key area for paleoclimate and paleoatmosphere research, since they are located in a transition zone between two precipitation belts, the extratropical westerlies receiving moisture from the Pacific and the tropical circulation with a continental/Atlantic moisture source.

In order to reconstruct past climate variations, especially those related to the El Niño Southern Oscillation (ENSO) phenomenon, a suitable glacier archive was sought. As one

[1] Paul Scherrer Institut, CH-5232 Villigen PSI, Switzerland; [2] Dept. für Chemie und Biochemie, Universität Bern, Freiestrasse 3, CH-3012 Bern, Switzerland; [3] Physikalisches Institut, Universität Bern, Sidlerstr. 5, CH 3012 Bern, Switzerland; [4] Geographisches Institut, Universität Bern, Hallerstr. 12, CH-3012 Bern, Switzerland; [5] VAW, Dept. Bau und Umwelt, ETH Zentrum, Gloriastrasse 37/39, CH-8092 Zürich, Switzerland; [6] Universidad de Chile, Departamento de Geografía, Marcoleta 250, Santiago de Chile, [7] Centro de Estudios Científicos (CECS), Arturo Prat 514, Valdivia, Chile; [8] FS Inventor, Muristr. 18, CH-3132 Riggisberg, Switzerland, [9] GSF-Institute for Hydrology, Neuherberg, D-85758 Oberschleissheim, Germany

* corresponding author: margit.schwikowski@psi.ch

The Patagonian Icefields: A Unique Natural Laboratory for Environmental and Climate Change Studies.
Edited by Gino Casassa et al., Kluwer Academic /Plenum Publishers, 2002.

possible candidate, the ice cap on the cerro Tapado was studied (5536 m, 30°08' S, 69°55' W, Region IV, Chile), which is located in the Norte Chico region of Chile,150 km east of La Serena in the Andean cordillera and on the border with Argentina (Figure 1). Because of its vicinity to the "South American Arid Diagonal" (Joussaume et al., 1986) the glacier is assumed to be affected by a discontinuous winter precipitation regime advected by westerlies. The major precipitation events therefore occur in the short period from May to August (Escobar, 1998) and are annually interrupted by a long dry period. The inter-annual variation in winter snow accumulation is influenced by the ENSO phenomenon, with larger snow accumulation in the Andean sector to the north of 35° S when the magnitude of the positive SST anomaly surpasses +1.0 °C during austral winter (Escobar and Aceituno, 1998).

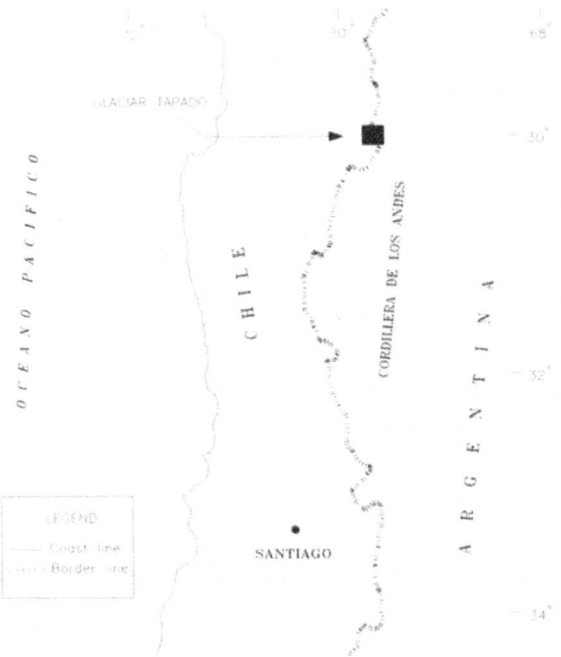

Figure 1. Location in Chile of cerro Tapado Glacier.

3. ICE-CORE RECOVERY AND CHEMICAL ANALYSES, METEOROLOGICAL AND GLACIOLOGICAL INVESTIGATIONS

In February 1999 we recovered a 36 m ice core reaching bedrock of the cerro Tapado ice cap (Figure 2). To do this, we used a new portable 77 mm electromechanical drill which was especially designed for employment on high-altitude glaciers (Ginot et al., in press; FS INVENTOR AG, Switzerland). Its total weight is only about 250 kg including power supply and sheltering tent and can be broken down into handy porter loads. Power is supplied by either flexible solar panels or a light-weight gasoline generator which are connected to an accumulator pack (Figure 3). The drill can be installed in less than one hour and in reasonable conditions it is possible to drill 50 m in 10 working hours. The recommended operation range is a maximum depth of 150 m. All parts in contact with the ice cores are composed of anodized aluminium and polyethylene. The recovered ice cores have a diameter of 77 mm and a maximal length of 900 mm.

Figure 2. View of the cerro Tapado ice cap.

Figure 3. The new portable 3-inch electromechanical drill installed in a sheltering tent and powered by flexible solar panels in operation on the cerro Tapado ice cap.

The ice cores were sealed in polyethylene tubes at the site and transported to the laboratory in frozen condition. All materials (vials, tubes, cutter) were carefully cleaned with ultra-pure water before coming in direct contact with the samples. Ice-core segments were photographed, measured, weighed, and cut in a cold room maintained at -20 °C. For the different chemical analyses, various cutting resolutions were applied. 1.5 cm sections of the inner part of the core were used to analyze concentrations of ionic species (Na^+, K^+, NH_4^+, Ca^{2+}, Mg^{2+}, F^-, $HCOO^-$, H_3CCOO^-, Cl^-, NO_3^-, SO_4^{2-}, $C_2O_4^{2-}$) by ion chromatography (Döscher *et al.*, 1995; Schwikowski, 1997). In 1.5 cm samples from the outer part of the core, stable isotope ratios ($\delta^{18}O$, δ^2H) were determined by mass spectrometry (see e.g. Schotterer *et al.*, 1997). Samples of 70 cm length were used to measure activities of tritium (see e.g. Schotterer *et al.*, 1998) and ^{210}Pb (Gäggeler *et al.*, 1983).

During the ice-core drilling campaign, surface snow experiments were performed in order to study the effect of sublimation and dry deposition on chemical and isotopical composition of the snow. For this purpose, 1 cm snow samples were collected from the surface twice daily over a 5 day period. In addition, samples were taken from a 38 cm snow pit with a resolution between 1 to 3 cm. In these snow samples, concentrations of major ionic species as well as stable isotope ratios $\delta^{18}O$ and δ^2H were determined. Sublimation was measured with the *lysimeter* technique (Hastenrath, 1997), and meteorological parameters such as temperature, relative humidity, wind direction and speed were recorded at the drilling site. This was accompanied by monitoring the same parameters at 4215 m on a moraine shoulder close to the glacier (Begert, 1999) and temperature at 5250 m on the ice margin since April 1998.

Detailed ice thickness measurements were carried out with a portable radio-echo sounding system. The ground-based digital impulse radar system used a transmitter developed by the University of Bristol, UK, with a maximum output voltage of 670 V_{pp}. The antennas consisted of resistively-loaded dipoles, with a 5 m antenna length, which result in a central frequency of about 10 MHz. The receiver is composed of a digital FLUKE oscilloscope connected to the receiving antennas, transferring the data through a serial port to a portable PC, where they were stored on the hard disk. The antennas were mounted on fiberglass fishing rods, with a separation of 10 m between the receiver and the transmitter. The whole system was carried by two persons, who walked on the surface of the glacier measuring one point every 5 seconds. In order to obtain a geographic position for each thickness measurement, a topographic-quality GPS receiver was used to fix the positions of the stakes. By means of differential correction methods, GPS data was obtained simultaneously at base camp and the horizontal precision attained was 5 m. In addition, firn temperature measurements were performed in the borehole using a thermistor chain.

4. RESULTS AND DISCUSSION

4.1. Glacier characteristics

In this Norte Chico area of Chile, the cerro Tapado appears to be the only summit having a true glacier flowing on its south-western slope. The adjacent mountains, even those of higher altitude such as the cerro de Olivares (6252 m), show only small remains of glaciers or snow fields on the summit or in the crater. The glacier on cerro Tapado extends from 5536 m down to its terminal moraine at 4500 m and is relatively flat at the top. Adjacent to the summit ice cap in the North and East are steep rock faces, free of ice cover. The transition between glacier and rock is characterized by the presence of several small frozen lakes and penitent fields, characteristic of significant ablation affecting the surface of the glacier. This is also observed at other glaciers located close to the "South American Arid Diagonal".

Three radar profiles were measured (Figure 4) indicating a maximum ice thickness of 42 m.

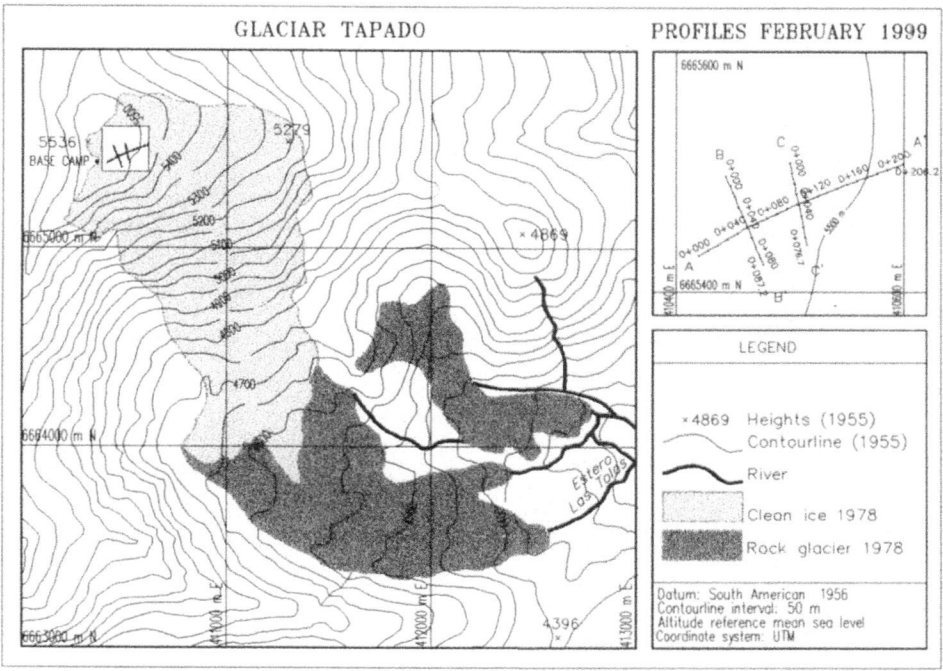

Figure 4. Topographic map of the glacier on the cerro Tapado along with the location of the radar profiles.

Near the ice-core drilling site the ice thickness reaches 33 m (Figure 5). This corroborates earlier results from the first radar survey conducted during the exploratory drilling of shallow firn cores in 1998. The ice thickness data was georeferenced to the regular cartography of the IGM (Instituto Geográfico Militar of Chile), which was compiled and digitized in order to produce the final chart of the study area (Figure 4).

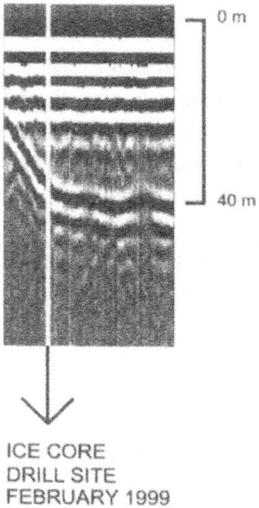

Figure 5. Radar reflectivities indicating ice thickness along profile A-A'

The firn temperatures decrease from -8.5 °C at the surface to -12.5 °C near bedrock (Figure 6), indicating that the glacier consists of cold firn and ice which is frozen to the ground. This presence of cold firn and ice is a prerequisite for the conservation of glaciochemical records of water-soluble species.

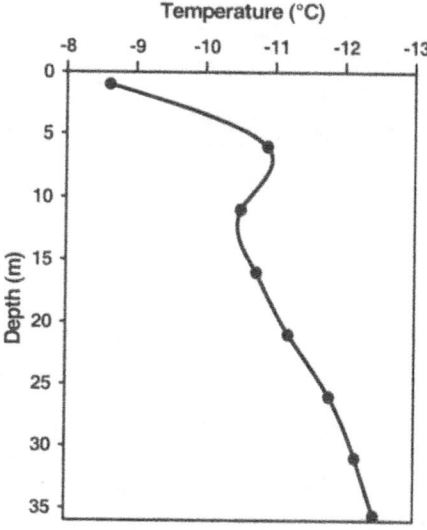

Figure 6. Firn temperatures measured in the 36 m borehole.

During the 8 days of direct measurements, a daily rate of sublimation of 1.89 mm water equivalent (weq) was observed, representing nearly 5 mm of snow. Sublimation showed a diurnal cycle with maximum values during the day in accordance with the diurnal variation of the meteorological parameters causing sublimation: atmospheric temperature and moisture, radiation and wind speed (Hastenrath, 1997; Cline, 1997). The obtained sublimation rate is significantly higher than the 0.1 mm reported for the period from morning to evening on Quelccaya ice cap (Hastenrath, 1978), but comparable to the 1.1 mm/day observed at Zongo Glacier during the dry season (Wagnon *et al.*, 1999). At El Laco in northern Chile (23.85° S, 67.49° W, 4400 m), located on the northern margin of the "South American Arid Diagonal", a maximum sublimation rate of 3 mm/day was estimated (Vuille, 1996).

The observed sublimation rate is corroborated by mass balance modeling (Kull, 1999), predicting an average sublimation of about 2 mm/day for the period between the last significant snowfall event in September 1998 and the beginning of the drilling campaign in February 1999. However, modeling results using the mean meteorological conditions in this region suggest that glaciation cannot exist due to the lack of moisture. An equilibrium line altitude (ELA) is not existent up to 6000 m and this is obviously the reason for the missing glaciation on adjacent mountains.

In order to obtain an ELA at 5300 m on cerro Tapado (similar to field observations), an annual accumulation of 750 mm at 5500 m must be assumed. Annual sublimation removes 490 mm, resulting in a net accumulation of 260 mm which is in good agreement with the value estimated from ice-core analyses (see below). Thus, it is obvious that at the cerro Tapado, local climatic and topographic effects such as blowing snow and moisture capturing on the west facing wall in combination with the high albedo of the snow surface lead locally to higher accumulation rates and preserve a perhaps relict glaciation.

4.2. Effects of post-depositional processes on snow composition

Results from the 5-day surface snow experiment demonstrate the strong influence of post-depositional processes such as sublimation of water and dry deposition on the chemical composition of the exposed snow surface. Three classes of ionic species could be identified by their different concentration development. Group 1 consisting of NH_4^+, $HCOO^-$, H_3CCOO^-, and H^+ showed a weak increase or decrease in concentration. This indicates that these species were reversibly deposited by snow and were subsequently released from the snow surface. Group 2 is formed by Cl^-, K^+, SO_4^{2-}, NO_3^-, MSA, and $C_2O_4^{2-}$. These ions showed a medium increase in concentration, which can be directly related to the enrichment resulting from the removal of water by sublimation. Thus, group 2 ions were irreversibly deposited and were not significantly released from the snow surface. Group 3 consists of Ca^{2+}, Mg^{2+}, Na^+, and F^- which showed a strong increase in concentration over the 5-day period, suggesting that besides sublimation, dry deposition of soil particles also affected the concentration.

These results agree in principle with the concentration profile observed in a 38 cm deep snow pit. The topmost 3.5 cm of the snow pit showed enhanced concentrations of group 2 and 3 ions compared to the lower part. From the enrichment, an exposure time of 120 to 160 days was estimated, during which sublimation and dry deposition affected the surface snow layer. This is consistent with meteorological measurements recording the last significant snowfall from 11 to 15 September 1998, i.e. about 150 days before the snow pit was sampled.

4.3 Glaciochemical record

Annual layer-counting using seasonal parameters cannot be applied throughout the core because of the absence of seasonal distribution of precipitation at the cerro Tapado (see above), which causes the lack of an annually-varying signal. Therefore, dating of the ice core was performed using the measured activity of the naturally produced radioactive isotope ^{210}Pb that decays with a half-life of 22.3 years. This method has been successfully used to establish the chronology of ice cores from Alpine Glaciers for time-scales of about 100 years (Gäggeler *et al.*, 1983; Eichler *et al.*, 2000). A linear regression fit of the ^{210}Pb activities reveals an age of about 100 years for the principal part of the core (Figure 7).

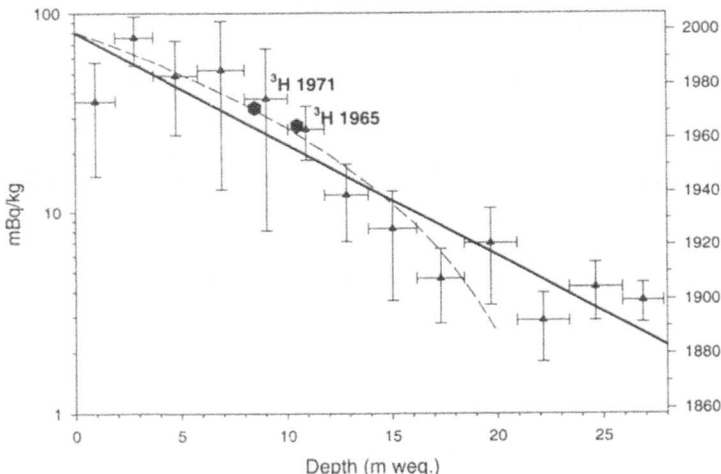

Figure 7. ^{210}Pb activities versus depth of the ice core together with a linear regression fit. In addition, the age-depth relationship obtained by a simple ice-flow model considering ice thinning is shown (dashed line). Two stratigraphic horizons are indicated. The right axis gives the resulting time scale. 1 Bq corresponds to 1 disintegration per second.

With this fit the thinning of the ice is not taken into account. The effect of thinning is illustrated by the depth-age relationship obtained with a simple 1D ice flow model. This agrees well with stratigraphic horizons such as the tritium maxima due to thermonuclear weapon tests, which occurred in 1965 and 1971 in the Southern Hemisphere. Applying this dating, gives a mean annual net accumulation of about 300 mm. Preliminary results using dust horizons as indicator to identify individual years in the record reveal a variation in annual accumulation which is in agreement with precipitation recorded at a nearby weather station (La Laguna). Higher annual accumulation and precipitation rates were observed during El Niño periods. The effect of the El Niño phenomenon on the accumulation at the cerro Tapado Glacier is additionally illustrated by good correlation with the Sea Surface Temperature Anomalies (SST) associated with El Niño for the period 1985-1999, for which annual layers could be identified in the core using dust horizons. In agreement with results from the surface snow experiments, distinct effects of sublimation and dry deposition were observed in the chemical records (Figure 8).

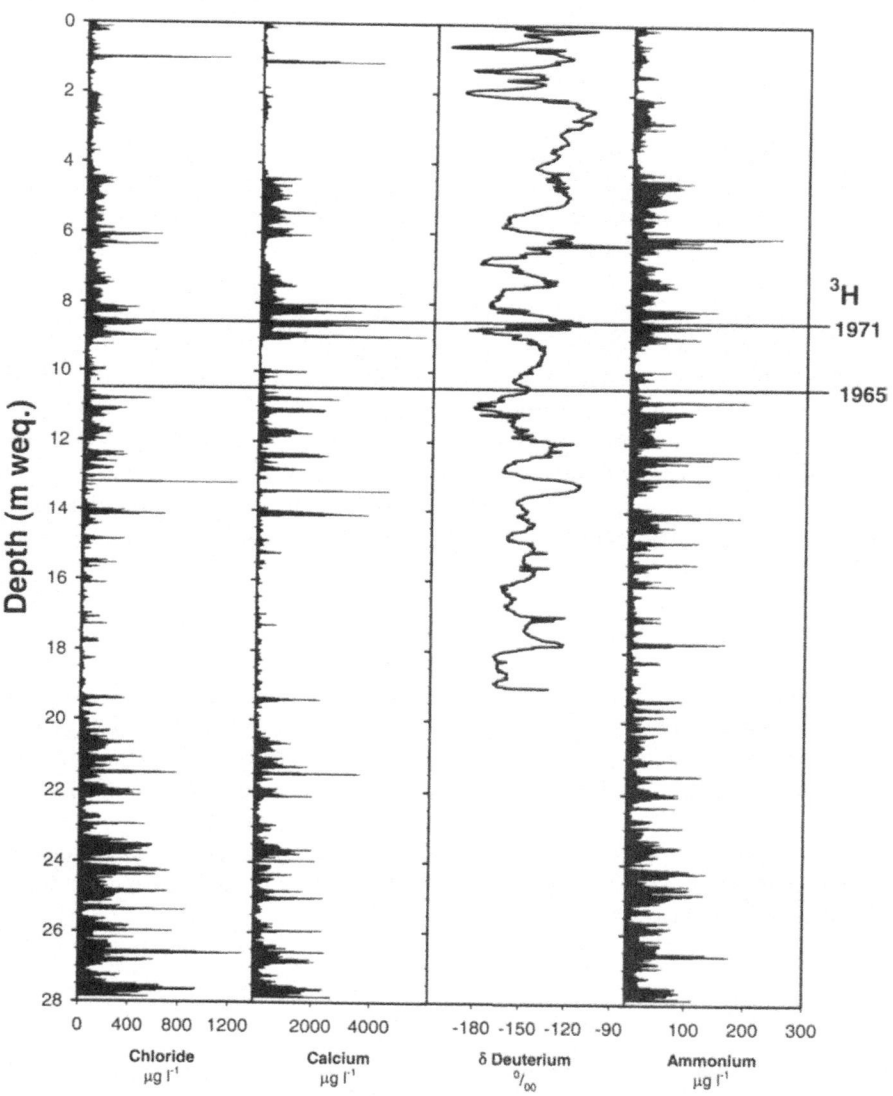

Figure 8. Concentrations of Cl⁻, Ca²⁺, and NH₄⁺ as well as the isotopic ratio δ^2H versus depth of the ice core (28 m weq represent 36 m absolute depth).

Strong sublimation is marked by concentration peaks of chemical species (e.g. Cl⁻, Ca^{2+}, NH_4^+, at 6 and 8 m weq. in Figure 8). Dry deposition resulted in horizons located just above the levels marked by sublimation, characterized by high concentrations of ionic species contained in the surrounding volcanic rocks (Ca^{2+}, Mg^{2+}, SO_4^{2-}, K^+). However, wet periods were characterized by low concentrations of chemical species (for example between 0 and 2 m weq. in Figure 8). Thus the variation in the concentration of chemical species in the core allows identification of dry and wet periods in this area that could be related to the La Niña/El Niño phenomenon.

5. CONCLUSIONS

Initial glaciochemical data retrieved from the 36 m cerro Tapado ice core indicate that a time period of about 100 years is accessible by this archive. The year-to-year variability of annual accumulation at this glacier site can strongly be influenced by the La Niña/El Niño phenomenon, with higher accumulation during El Niño periods. This is consistent with precipitation data recorded at a nearby meteorological station in the Elqui Valley.

Due to the fact that precipitation occurs mainly in the short period from May to August, post-depositional processes such as sublimation and dry deposition become important during the long dry period. As demonstrated by short-term snow-surface experiments, the effects of these processes on the chemical composition of the surface snow contribute significantly to the variability in the glaciochemical record. Dry and wet periods in this area related to the La Niña/El Niño phenomenon might be identified by enhanced and low concentrations of chemical species, respectively.

6. ACKNOWLEDGMENTS

We would like to thank Jorge Quinteros (Dirección General de Aguas), Alvaro Giannini (Universidad de Chile, Santiago), Benjamin Zweifel (ETH, Zürich), Bernard Pouyaud and Robert Gallaire (IRD, Bolivia) for their contribution to the field work. Appreciation is expressed to the Chilean Army (Regimiento de Infantería N° 21 Arica) for the support of the fieldwork, to Swisscargo for transporting the ice samples and to Argauer Zentralmolkerei for storing the ice. This work was carried out in the frame of the Swiss National Science Foundation, Project # 21-50854.97. The Chilean author was supported by project Fondecyt 1980293.

7. REFERENCES

Begert, M., 1999, Klimatologische Untersuchungen in der weiteren Umgebung des Cerro Tapado, Norte Chico, Chile, Diploma Thesis, Geographical Institute, University of Bern (in German).
Cline, D. W., 1997, Snow surface energy exchanges and snowmelt at a continental mid-latitude alpine site, *Water Resources Research*, 33:689-701.
Döscher, A., Schwikowski, M., and Gäggeler, H. W., 1995, Cation trace analysis of snow and firn samples from high-alpine sites by ion chromatography, *Journal of Chromatography*, A 706:249-252.

Eichler, A., Schwikowski, M., Gäggeler, H. W., Furrer, V., Synal, H.-A., Beer, J., Sauer, M., and Funk, M., 2000, Glaciochemical dating of an ice core from the upper Grenzgletscher (4200 m a.s.l.), *Journal of Glaciology,* **46**:507-515.

Escobar, F., 1998, A 20 year daily precipitation and temperature data set measured in 4 Chilean stations: Juntas, La Serena, La Laguna, Rivadavia, Dirección General de Aguas, Ministerio de Obras Públicas, Morandé 59, Santiago, Chile.

Escobar, F., and Aceituno, P., 1998, The ENSO phenomenon influence on snow fall in the Andean sector of central Chile during winter, *Bulletin Institut Francés d'Études Andines,* **27**:753-759.

Gäggeler, H., von Gunten, H. R., Rössler, E., Oeschger, H., and Schotterer, U., 1983, ^{210}Pb-Dating of cold alpine firn/ice cores from Colle Gnifetti, Switzerland, *Journal of Glaciology,* **29**:165-177.

Ginot, P., Stampfli, F., Stampfli, D., Schwikowski, M., and Gäggeler, H. W., in press, FELICS, a new ice core drilling system for high altitude glaciers, in: Proceedings of the International Workshop "Ice Drilling Technology", Nagaoka, Niigata, Japan, 30 October-1 November 2000, Memoirs of National Institute of Polar Research, Special Issue, **56**.

Hastenrath, S., 1978, Heat-budget measurements on the Quelccaya ice cap, Peruvian Andes, *Journal of Glaciology,* **20**:85-97.

Hastenrath, S., 1997, Measurement of diurnal heat exchanges on the Quelccaya ice cap, Peruvian Andes, *Meteorology and Atmospheric Physics,* **62**:71-78.

Joussaume, S., Sadourny R., and Vignal, C., 1986, Origin of precipitating water in a numerical simulation of July climate, *Ocean-Air Interactions,* **1**:43-56.

Kull C., 1999, Modellierung paläoklimatischer Verhältnisse basierend auf der jung-pleistozänen Vergletscherung in Nordchile - Ein Fallbeispiel aus den Nordchilenische Anden, *Zeitschrift für Gletscherkunde und Glazialgeologie,* **35**:3 5-64 (in German).

Schotterer, U., Froehlich, K., Gäggeler, H. W., Sandjordj, S., and Stichler, W., 1997, Isotope records from Mongolian and Alpine ice cores as climatic indicators, *Climatic Change,* **36**:519-530.

Schotterer, U., Schwarz, P., and Rajner, V., 1998, From the prebomb levels to industrial times. A complete tritium record from an alpine ice core and its relevance for environmental studies, in: *Isotope Techniques in the Study of Environmental Change,* Proceedings of International Symposium on Isotope Techniques in the Study of Past and Current Environmental Changes in the Hydrosphere and the Atmosphere, Vienna, 14-18 April 1997, IAEA.

Schwikowski, M., 1997, Analytical chemistry in high-alpine environmental research, *Chimia,* **51**:8-13.

Vuille, M., 1996, Zur raumzeitlichen Dynamik von Schneefall und Ausaperung im Bereich des südlichen Altiplano, Südamerika, Ph.D. thesis, Geographical Institute, University of Bern (in German).

Wagnon, P., Ribstein, P., Kaser, G., and Berton, P., 1999, Energy balance and runoff seasonality of a Bolvian glacier, *Global and Planetary Change,* **22**:4-58.

INTERPRETING CLIMATE SIGNALS FROM A SHALLOW EQUATORIAL CORE: ANTISANA, ECUADOR

Mark W. Williams[1,2*], Bernard Francou[3], Eran Hood[1,2], and Bruce Vaughn[2]

1. ABSTRACT

The potential use of equatorial glaciers to record past climates and precipitation sources in ice cores is unknown. A shallow (16-meter) core was recovered from the summit of Antisana in November of 1999 to evaluate whether it was worthwhile to drill a full-depth ice core with more rigorous post-collection processing. Ice lenses were found in the top 200 cm, all with a thickness less than 1 cm, suggesting occasional melt at the snow surface but little redistribution of water isotopes by percolating liquid water. Density in the core below the seasonal snow accumulation ranged from 450-720 kg m^{-3}, indicating that we sampled the firn layer but not ice. The ^{18}O content of the core ranged from a minimum value of -23.4‰ to a maximum value of -9.9‰. There appeared to be an oscillating signal in the isotopic content of the firn core. Periodic maxima of about -10 to -12‰ occurred at depths of 80, 510, 860, 1300, and 1520 cm. The range and oscillations in the ^{18}O values are encouraging and suggest that additional effort is warranted to investigate the possible use of water isotopes to date equatorial ice cores. The similar values in slope between our local meteoric water line (8.1) and the global meteoric water line (8.0) suggest an absence of complex kinetic fractionation processes affecting precipitation on the summit of Antisana. However, the enriched deuterium excess value of 12 at Antisana compared to the global mean of 10 suggests that some of the water vapor that formed precipitation was derived from evaporation of localized water sources, such as the Amazon basin.

[1] Department of Geography, University of Colorado, Boulder, U.S.A.; [2] Institute of Arctic and Alpine Research (INSTAAR), University of Colorado, Boulder, U.S.A.; [3] Institut de Recherche pour le Developpement (France), Quito, Ecuador.

* corresponding author: markw@snobear.colorado.edu

The Patagonian Icefields: A Unique Natural Laboratory for Environmental and Climate Change Studies.
Edited by Gino Casassa et al., Kluwer Academic /Plenum Publishers, 2002.

2. INTRODUCTION

Tropical glaciers have been poorly studied, in part because they are generally located in countries with few resources and little need to study the glaciers. One of the best-studied systems of tropical glaciers is located on Mount Kenya in Africa. Glaciers on Mount Kenya have retreated dramatically between 1963 and 1987, possibly as a result of greenhouse forcing (Hastenrath and Kruss, 1992). In South America, there may be a strong connection between El Niño-Southern Oscillation (ENSO) events and glacial mass balances of tropical glaciers. ENSO events have been recorded in the stratigraphy of the Quelccaya Ice Cap in Peru (Thompson *et al.*, 1984). More recent work has shown that there is a negative correlation between ENSO events and the glacial mass balance of glaciers in both Peru and Bolivia (Francou *et al.*, 1995). These small changes in climate have the capacity to cause large changes in glacial hydrology (Ribstein *et al.*, 1995).

The increasing population pressure in Ecuador and other tropical mountain countries of South America has resulted in some urgency in understanding tropical glaciers and how these glaciers may respond to changes in climate. Mudslides from glaciers on Nevado Ruiz in Colombia killed 25,000 people in 1985 and glacial avalanches from Huascarán in Peru have killed more than 20,000 people in 1962 and 1970. Quito, the capital of Ecuador, now gets significant amounts of municipal and industrial water from the Antisana volcano's glacial runoff. Additionally, glacial runoff from Chimborazo, Cayambe, Cotopaxi, Carihuayrazo, and Altar, provide water for irrigation and domestic use for local communities. The success of the Antisana water diversion has resulted in proposals to expand existing uses of glacial meltwater and a call for additional water diversions and hydroelectric plants in Ecuador using glacial runoff.

There is reason to suspect that Ecuadorian glaciers are very sensitive to climatic change. Glacier #15 of Antisana Volcano has retreated 150 meters between 1994 and 1997, most likely because of increases in air temperature. How glacial runoff in Ecuador may respond to changes in climate is unknown. It is possible that with a 2-3 °C increase in global temperature, there would be massive glacial retreat throughout Ecuador. There may be a short-term increase in water availability as glaciers melt, followed by a dramatic decrease in water availability when glaciers melt-out completely.

Ice- and firn-core records have the potential to provide proxy climate information that may yield information on the causes of this glacier retreat. Confounding the interpretation of the climate signal in inner tropical ice cores is the lack of seasonal changes in temperature and precipitation, which drive isotopic and chemical differences. Here we report on the results of a 16-meter firn core drilled on the summit of Antisana, Ecuador (0.481° S 78.141° W) in November 1999, at an altitude of 5752 meters. Our primary objectives were to:

i.   evaluate whether there was sufficient range in the values of water isotopes such that a deep ice core could provide the potential for dating and determining precipitation sources.

ii.   evaluate whether there was a periodic signal in the water isotopes that may correspond to annual precipitation changes that could be used to date the core.

iii.   evaluate whether there was a change in water isotopes that may correlate with El Niño and La Niña events.

3. SITE DESCRIPTION

Antisana is an unusual combination of an active glacial system located on an active volcano at the equator. Antisana is 5752 m (18,871 ft) in elevation and located at 0°28'30" S latitude and 78°08'55" W longitude. The last eruption was in 1802; Antisana is still considered active. Antisana rises directly from the Amazon basin and forms the crest of the continental divide that separates drainage systems that flow into the Pacific Ocean to the west and the Atlantic Ocean to the east. The glacial hydrology of Antisana is of particular interest because glacial runoff supplies high-quality drinking water to Quito by two artificial collectors, one situated at Papallacta (North), and the other at the new reservoir La Mica (South).

The shallow firn core on Antisana was part of a larger effort to understand the dynamics of glacial systems in South America. The Institut de Recherche pour le Developpement (France) has been actively monitoring glaciers in the tropical Andes since 1991. The glaciers on Antisana are the only intensively monitored glaciers in the equatorial Andes.

The climate of Antisana is equatorial and wet, with precipitation occurring year round. Annual precipitation amounts to 1000 mm (measured) in the lower reaches of the glacierized area on the west side. However, interannual variability may be as high as 40%. The amount of annual precipitation at the summit is unknown, but it is most likely greater than 2000 mm. There is no evidence of a clear seasonality trend in the precipitation and the most rainy month may occur at any time from February to October. Only November-January includes a period of decreasing precipitation, which corresponds to the veranillo or little summer. Temperature is almost constant during the year, but the interannual variability is significantly high, being about 2 °C or even more. The 0 °C isotherm is close to 5000 m asl. Liquid precipitation is possible at elevations less than 5000 m asl. The glacier is probably polythermal, with a cold surface limited to the summit.

4. METHODS

The firn core was collected over a three-day period from 2 to 6 November 1999 at the summit of Antisana. A snow pit was excavated to a depth of 200 cm, at which time the snow became too hard for further excavation. A sidewall of the snow pit was analyzed for physical and chemical properties, following the protocols in Williams et al. (1996, 1999). A portable ice drill made by PICO (Polar Ice Coring Office) was then utilized to collect samples from 200 cm to 1600 cm. The general length of cores retrieved using the coring device was 0.5 m. The core was then sliced into sections of 10 to 20 cm in length, weighed, placed into plastic bags, and stored in a snow cave. Snow and ice samples were returned to Quito, melted, prepped for chemical and isotopic analyses, and immediately flown back to the United States.

Oxygen isotopes were analyzed using the CO_2 equilibration method of Epstein and Mayeda (1953). Values for the oxygen-18 (^{18}O) isotope of oxygen in the CO_2 gas were analyzed by mass spectrometry at the Institute of Arctic and Alpine Research in Boulder, CO. The deuterium (D) isotope of hydrogen (H) in water was analyzed at the Institute of Arctic and Alpine Research following the protocol in Vaughn et al. (1998). The ^{18}O and

D values are expressed in the conventional delta (δ) notation as the per mil (‰) difference relative to the international Vienna Standard Mean Ocean Water (VSMOW):

$$\delta^{18}O = \frac{{}^{18}O/{}^{16}O_{sample} - {}^{18}O/{}^{16}O_{standard}}{{}^{18}O/{}^{16}O_{standard}} \times 1000$$

$$\delta D = \frac{D/H_{sample} - D/H_{standard}}{D/H_{standard}} \times 1000$$

5. RESULTS AND DISCUSSION

The ^{18}O content of the ice ranged from a minimum value of -23.4‰ to a maximum value of -9.9‰ (Figure 1). The range of 13.5‰ was quite surprising. Moreover, changes in the ^{18}O values occurred very rapidly, from -9.9‰ at a depth of 80 cm to -23.4‰ at a depth of 200 cm. The large range in the ^{18}O values indicates that water isotopes may provide information that is helpful for dating ice cores and for evaluating sources of precipitation.

There appears to be an oscillating signal in the isotopic content of the firn core. Periodic maxima of about -10 to -12‰ occurred at depths of 80, 510, 860, 1300, and 1520 cm. Whether this oscillation in the isotopic content of the firn core represents an annual signal is unknown. The oscillations in the ^{18}O values are encouraging and suggest that additional effort is warranted to investigate the possible use of water isotopes to date equatorial cores.

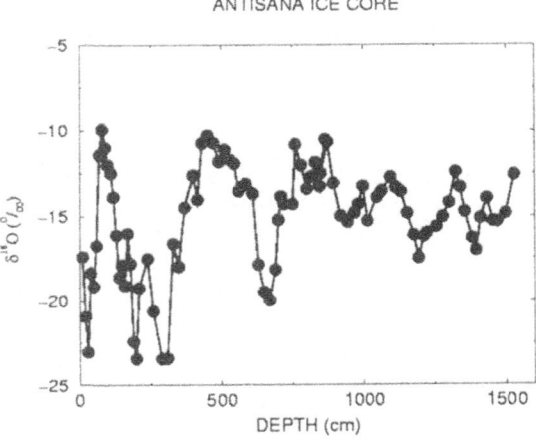

ANTISANA ICE CORE

Figure 1. The isotopic content of ^{18}O as a function of firn-depth on Antisana. The shallow firn core was collected using a hand-drill.

Empirical results have shown that hydrogen and isotopic values in precipitation co-vary and are generally described by the relationship (Craig, 1961),

$$\delta D = 8\ \delta^{18}O + 10\ (\text{‰})$$

which is defined as the global meteoric water line. Deuterium values in the Antisana firn core were highly correlated with ^{18}O values ($r^2 = 0.99$, $n = 88$; Figure 2). Our local meteoric water line has a similar slope but a different y-intercept (or deuterium excess value):

$$\delta D = 8.\ 1\ \delta^{18}O + 12.8\ (\text{‰})$$

Deuterium excess (d) was calculated following the protocol developed by Johnsen and White (1989) based on the equation for the Global Meteoric Water Line (Craig, 1961):

$$d = \delta D\text{-}8\ \delta^{18}O$$

The similar values in slope between our results (8.1) and the global meteoric water line (8.0) suggests an absence of complex kinetic fractionation processes affecting precipitation on the summit of Antisana. However, the enriched deuterium excess value of 12.8 at Antisana compared to the global mean of 10, suggests that some of the water vapor that formed precipitation was derived from evaporation of localized water sources.

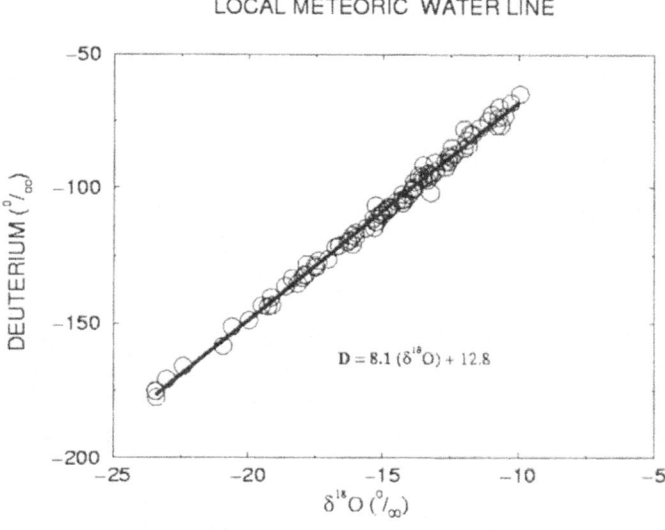

Figure 2. Local meteoric water line from the Antisana firn core. The n=88 sample points are roughly equally spaced over the entire 15.2-m range of the core, as described in 4. Methods.

A closer look at the deuterium excess values from the Antisana firn core shows that they ranged from 8.6 to 18.1 (Figure 3). The higher deuterium excess values of around +15 are correlated with enriched ^{18}O values of about -10‰. The elevated deuterium excess values and enriched ^{18}O values are consistent with a source of moisture that has undergone evaporation. Recycled water vapor from the Amazon basin that originated from the Atlantic Ocean may be indicated by the elevated values for deuterium excess and ^{18}O. The lower deuterium excess values and ^{18}O values are consistent with equilibrium reactions and indicate a Pacific Ocean source.

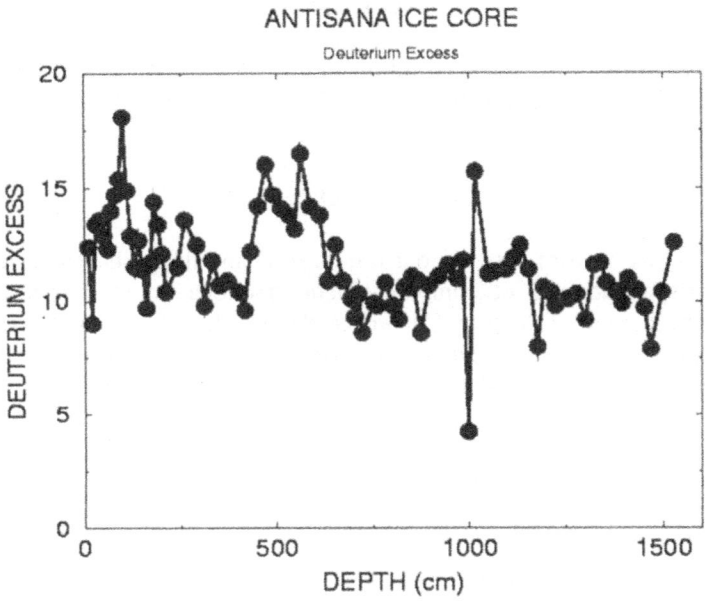

Figure 3. Deuterium excess as a function of firn-depth on Antisana.

One hypothesis for the trends in the $\delta^{18}O$ is that the variations reflect seasonality in moisture sources rather than differences in temperature. For example, migration of the Intertropical Convergence Zone (ITCZ) could provide changing sources of cloud moisture. The dampening of the isotopic signal at depth may be caused by isotopic diffusion. However, if accumulation rates are in the order of 2 meters per year, diffusion is probably not an important process at this site.

The alternative hypothesis is that the heavier peaks in the $\delta^{18}O$ record at depths of 100 and 400 cm may be related to changes in moisture sources caused by ENSO events in the last couple of years. Without adequate depth-age constraints, demonstrating absolute connections between ENSO events and isotopic signatures remains difficult. In either case, it appears that the trend of the base line shifts to lighter values in the top 4 meters.

6. ACKNOWLEDGMENTS

We greatly appreciate the logistic support provided by Instituto Nacional de Meteorología e Hidrología (Ecuador) and Empresa Municipal de Alcantarillado y Agua Potable de Quito (Ecuador).

Funding was provided in part by a Fulbright Fellowship, a Faculty Fellowship from the University of Colorado, Boulder, USA, a UROP award from CU-Boulder, USA and several programs within the National Science Foundation, including International Programs, Hydrology, Long Term Ecological Research, and GRT-International.

7. REFERENCES

Craig, H., 1961, Isotopic variations in meteoric waters, *Science,* **133**:1702-1703.

Epstein, S., and Mayeda, T., 1953, Variations of the ^{18}O content of waters from natural sources, *Geochimica et Cosmochimica Acta,* **4**:213-224.

Francou, B., Ribstein, P., Sémiond, H., and Rodríquez. A., 1995, Balances de glaciares y clima en Bolivia y Perú: impactos de los eventos ENSO, *Bulletin Institut Francés d'Études Andines,* **24**:661-670.

Hastenrath, S., and Kruss, P. D., 1992, The dramatic retreat of Mount Kenya's glaciers between 1963 and 1987: greenhouse forcing, *Annals of Glaciology,* **16**:127-134.

Johnsen, S. J., and White, J. W. C., 1989, The origin of Arctic precipitation under present and glacial conditions, *Tellus,* **41B**:452-468.

Ribstein, P., Tiriau, E., Francou, B., and Saravia, R., 1995, Tropical climate and glacier hydrology, *Journal of Hydrology,* **165**:221:234.

Thompson, L. G., Mosley-Thompson, E., and Arnao, B. M., 1984, El Niño-Southern Oscillation events recorded in the stratigraphy of the tropical Quelccaya Ice Cap, Peru, *Science,* **226**:50-53.

Vaughn, B. H., White, J. W. C., Delmotte, M., Trolier, M., Cattani, O., and Stievenard, M., 1998, An automated system for hydrogen analysis of water, *Chemical Geology,* **152**:309-319

Williams, M. W., Brooks, P. D., Mosier, A., and Tonnessen, K. A., 1996, Mineral nitrogen transformations in and under seasonal snow in a high-elevation catchment, Rocky Mountains, USA, *Water Resources Research,* **32**:3175-3185.

Williams, M. W., Cline, D., Hartman, M., and Bardsley, T., 1999, Data for snowmelt model development, calibration, and verification at an alpine site, Colorado Front Range, *Water Resources Research,* **35**:3205-3209.

POTENTIAL APPLICATION OF METHODOLOGIES FOR STUDYING ATMOSPHERIC AEROSOLS, IN THE STUDY OF CLIMATIC AND ENVIRONMENTAL INFORMATION FROM ICE CORES

Margarita Préndez[1*]

1. ABSTRACT

This paper presents our most important work on atmospheric aerosols from long-term aerosol studies during the last twenty years in the South Shetland Islands, Antarctic Peninsula.

The objective is to study the potential for extended use of existing methodologies to measure both the particulate matter in the Antarctic atmosphere and to perform future analyses of climatic and environmental information from ice cores.

2. INTRODUCTION

The atmosphere contains many components, which can be natural or man-made gases and suspended particulate matter, called particles or atmospheric aerosols, which are specifically the small particles or fine droplets of liquid suspended in the air. These can be of different sizes, forms and chemical composition.

Antarctica is a particularly good region for atmospheric chemistry studies because it has been little affected by human activity. However, over the last few years, human activity has increased. In order to protect the Antarctic ecosystems, an international treaty, the Madrid Protocol, was established to regulate activities in the Antarctic environment. Chile ratified the Protocol in February 1998.

Our group has been working on the Antarctic Peninsula since 1980. Our aim has been to analyze and describe the present atmosphere, as well as to search for the cleanest

---

[1] Laboratorio de Química de la Atmósfera, Facultad de Ciencias Químicas y Farmacéuticas, Universidad de Chile, Casilla 233, Santiago, Chile.

* corresponding author: mprendez@ll.ciq.uchile.cl

The Patagonian Icefields: A Unique Natural Laboratory for Environmental and Climate Change Studies
Edited by Gino Casassa et al., Kluwer Academic /Plenum Publishers, 2002.

place on the Peninsula as a base reference for data measurements, which is important to know in the new context of the Madrid Protocol.

Our research plan has a two-fold purpose: to discover which natural components are present in the particulate matter and to try to understand how much of an increase in some components is acceptable. This knowledge is essential for establishing appropriate regulations in the control of atmospheric pollution.

The work presented here concentrates on our findings from one of the three study locations, King George Island, which is of easy access, on the north tip of the Antarctic Peninsula. This island has the largest human population of all of the South Shetland Islands and at the present time, scientific stations from many different countries occupy the island.

We have participated in summer campaigns in 1980, 87, 88, 89, and 1995-99 on King George Island. Using a so-called Cascade Impactor, we were able to measure the total chemical composition of the aerosols, as well as their chemical composition relative to a definite particle size (particle diameter: 2.84, 2.04, 1.4, 0.8, 0.41 and <0.41 µm). A long-term study will enable us to evaluate the impact of an increasing human population on the island.

Since 1996, we have been performing year-round campaigns. A different instrument, Partisol 2000, has been used to collect total atmospheric aerosols with a particle diameter smaller than 10 µm.

3. ANALYTICAL TECHNIQUES

The first objective is to discover which inorganic species - elements and ions - and which organic compounds are present in the particles. As in general the quantities of particulate matter collected are very small (milligram concentrations) it is necessary to use very sensitive analytical instrumentation and clean rooms for sample preparation.

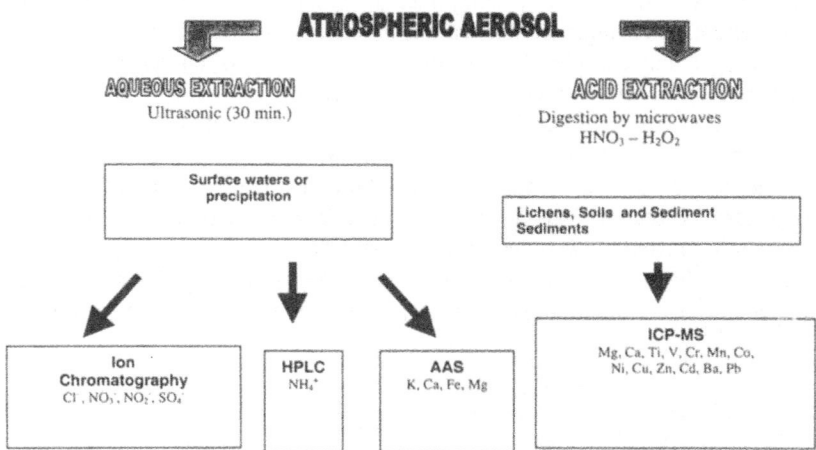

Figure 1. Schematic procedure used to prepare samples of particulate matter to quantify ions and elements by the following analytical techniques: Ion Chromatography (IC), High Pressure Liquid Chromatography (HPLC), Atomic Absorption Spectrophotometry (AAS), Inductively Coupled Plasma - Mass Spectrometry (ICP-MS).

One of the main problems experienced by atmospheric chemists, is difficulty in correctly determining the origin and source of the different chemical species that are found in particulate matter. Figure 1 is a schematic diagram of the particulate matter sampling procedure for subsequent ion and element quantification using the different techniques described here.

Table 1 shows the corresponding detection limits for the different techniques. The detection limits estimated in Table 1 are approximate values because they depend critically on the exact chemical species to be quantified.

Table 1. Analytical instrumentation to quantify ions and elements in the particulate matter and detection limits for the different techniques used to quantify elements (AAS, ICP-MS), ions (HPLC, CE, IC) and chemical compounds (GC-MS).

	Quantification	Detection limits*
	AAS	≥ 0.5 mg/L
	ICP-MS	≥ 1 µg/L
	HPLC	≥ 10 µg/L
	CE	≥ 10 µg/L
	IC	≥ 1 µg/L
	GC-MS	≥ 10 µg/L

Ion Chromatography (IC), High Pressure Liquid Chromatography (HPLC), Atomic Absorption Spectrophotometry (AAS), Inductively Coupled Plasma - Mass Spectrometry (ICP-MS), Capillary Electrophoresis (CE) and Gas Chromatography - Mass Spectrometry (GC-MS).
*All values are estimated values because the detection limits depend on the exact chemical species to be quantified.

Table 2 shows, as an example, detailed element detection limits for 30 elements quantified by ICP-MS.

Table 2. Detection limits (µg/L), for 30 of the 64 elements that can be quantified by ICP-MS.

Element	Detection limit (µg/L)	Element	Detection limit (µg/L)
Lithium	1.00	Tin	0.10
Beryllium	0.04	Antimony	0.04
Boron	1.00	Tellurium	0.10
Sodium	4.00	Cesium	0.05
Magnesium	0.10	Barium	0.05
Aluminum	0.40	Lanthanum	0.04
Strontium	0.06	Cerium	0.04
Scandium	0.10	Neodymium	0.04
Titanium	0.20	Copper	0.20
Vanadium	0.20	Arsenic	0.40
Chromium	0.30	Rubidium	0.20
Manganese	0.10	Molybdenum	0.07
Cobalt	0.04	Silver	0.20
Nickel	0.30	Cadmium	0.06
Zinc	0.60	Lead	0.20

Many methodologies exist for determination of the origin and sources of the different chemical species found in particulate matter, for example statistical methods or those involving meteorological parameters. However, apparently straightforward methods are complicated by many factors, such as the existence of numerous biogeochemical cycles and other known and unknown interrelated factors. Thus, we have discovered that in order to obtain good results, it is also necessary to evaluate the same chemical species studied in aerosols in other parts of the environment. Therefore, since 1996, , we have also been evaluating precipitation, soil, continental waters and the impact of atmospheric pollutants on lichen biomonitors, as well as monitoring particles

4. RESULTS AND DISCUSSION

Figure 2 compares PM-10 particulate matter (diameter smaller than 10 μm) collected from March to June of 1996, 1997 and 1998 on King George Island. The PM-10 concentration is very small compared to values corresponding to urban sites elsewhere on Earth. However, monthly and annual variations can be seen (Troncoso and Préndez, 1999), due essentially to changing meteorological factors such as wind direction and speed, and a combination of precipitation and marine aerosols.

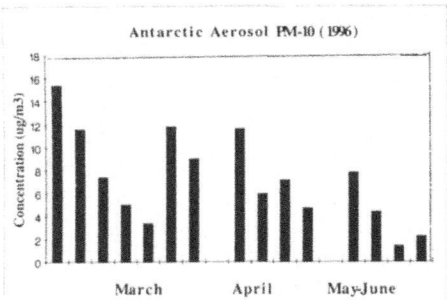
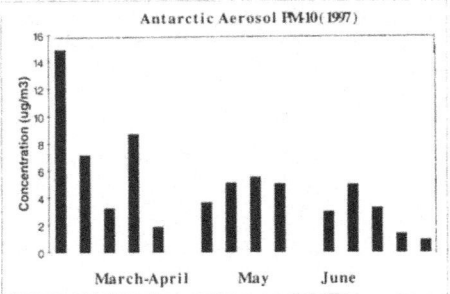
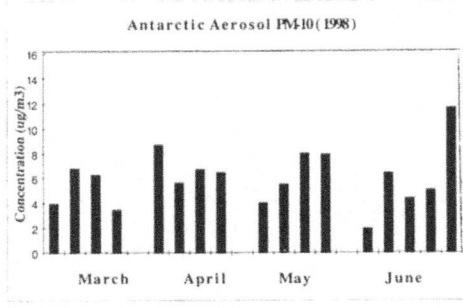

Figure 2. Concentration of PM-10 particulate matter (diameter smaller than 10 μm) collected from March to June of 1996, 1997 and 1998 on King George Island, South Shetland Islands, Antarctic Peninsula. The bars represent individual samples taken at different times within each month. Troncoso and Préndez, 1999. Revista Serie Científica INACH **49**:132-134. Reproduced with kind permission of the Instituto Antártico Chileno.

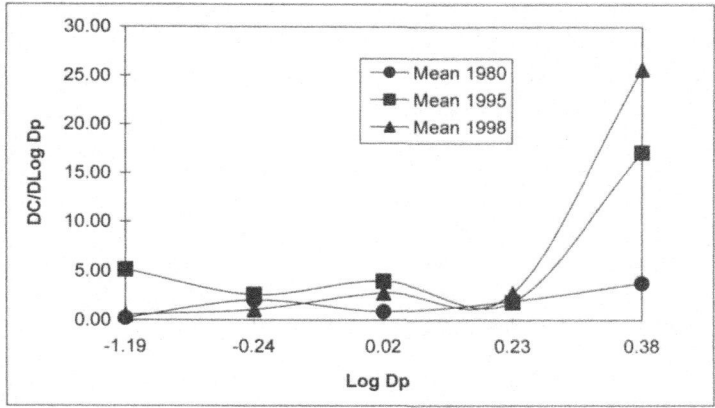

Figure 3. Mean mass/size distribution versus log base 10 of size diameter of PM-3 particulate matter (diameter smaller than 3 μm), collected during summertime of 1980, 1995 and 1998 at King George Island.

Figure 3 shows the mean values of mass/size distribution of PM-3 particulate matter (diameter smaller than 3 μm) versus particle size diameter during the summertime of three different years, 1980, 1995 and 1998, on King George Island. The y-axis corresponds to a parameter correlated to a relative concentration of the different fractions of PM-3 and the x-axis corresponds to the log of the particle diameter in PM-3 fractions (adapted from Préndez and Alcota, 1999).

Results show that the relative contribution of different particle sizes increases in the period 1980-1998, for larger particles. This is probably due to greater human influence; more soil particles are suspended in the air, produced by increased activity on the island, especially heavier vehicle traffic, driven by an increasing population. This assumption was confirmed by analysis of similar curves for a range of elements that are normally found in the soil of the island.

Table 3 shows the mean soil concentration of certain elements on King George Island. Note that the island's soil element composition is quite different from that of the terrestrial crust. Some elements, copper (Cu) and cadmium (Cd) are much more enriched in the soil, while others, such as calcium (Ca), Chromium (Cr), Nickel (Ni) and Lead (Pb) are more enriched in the terrestrial crust.

Table 3. Mean element concentration in soil on King George Island and in the terrestrial crust, expressed as units of element mass/total mass.

Major Elements	King George Island % of concentration	Terrestrial Crust* % of concentration
Fe	5.54 ± 1.0**	5.6
Ca	2.5 ± 0.1***	3.6
Na	2.1 ± 0.2***	2.8
Mg	1.8 ± 0.1	2.1

Table 3. Mean element concentration in soil on King George Island and in the terrestrial crust, expressed as units of element mass/total mass (continued).

Trace Elements	King George Island Conc. (mg/kg)	Terrestrial Crust* Conc. (mg/kg)
Mn	1201 ± 156**	950
Cr	30 ± 11**	100
Ni	44 ± 3.5**	75
Zn	81 ± 20**	70
Cu	129 ± 13.0***	55
Co	32.7 ± 4.4***	25
Pb	5.6 ± 0.8***	12.5
Cd	5.6 ± 1.2**	0.2

* From Mason, 1966.
** Values from Carrasco and Préndez, 1991. Errors taken from variation coefficient expressed as %.
***From Aponte, 1998.

Figure 4 shows the concentration of anions (chlorides, sulfates and nitrates), in the PM-10 particulate matter collected from January to September 1996 (Aponte, 1998). There is a clear decrease in concentration of all anions, which may be related to changing meteorological conditions such as wind speed, direction, precipitation and also changes in the source of such anions. It is well known, for instance that the concentration and chemical composition of particulate matter is affected by wet precipitation, as opposed to solid precipitation, because some chemical species in aerosols are easily dissolved by rain due to increased solubility of anions in water. Concentration of aerosol anions could therefore be used as a proxy for detecting climate changes.

Figure 5 shows the mean annual concentrations of several specific anions and elements in Antarctic precipitation for three different years (1995, 1996, and 1999). It is clear that there is annual variation and differences in the relative contribution of the species under study, which would certainly be reflected in surface waters and snow. The proximity of the sampling site to the ocean and the strong winds affecting the area could explain the greater concentrations of chloride compared to sulfates or nitrates. However, the absence of a correlation between chloride and sodium or calcium, implies a more complex explanation is required for chemical composition behavior during the three years studied.

It is interesting to note that the nitrate is almost exclusively of atmospheric origin (Kerminen *et al.*, 2000.), while the presence of trace elements is much more closely associated with human activity, a simple fact that could explain the annual variations. It has been noted that some man-made elements come from a local source, while others are from long distance sources (Boutron and Wolf, 1989). This is the case for lead (Pb), which has also been detected in aerosols and in lichen biomonitors (Préndez *et al*, 1999).

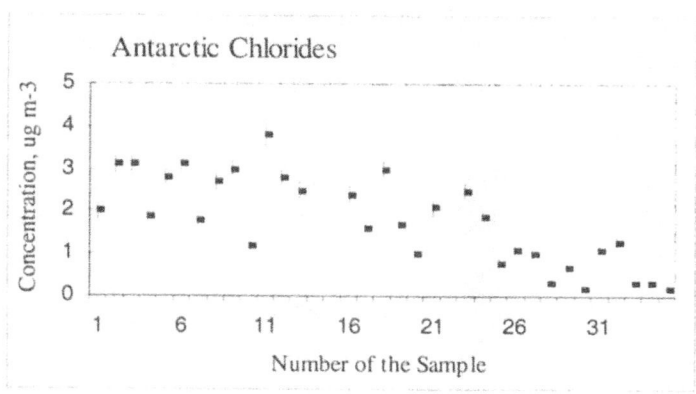

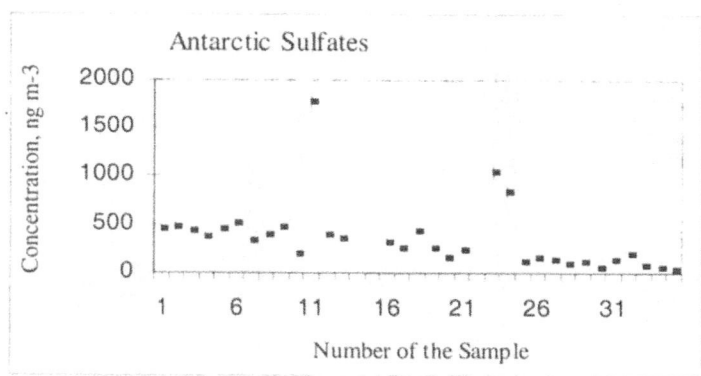

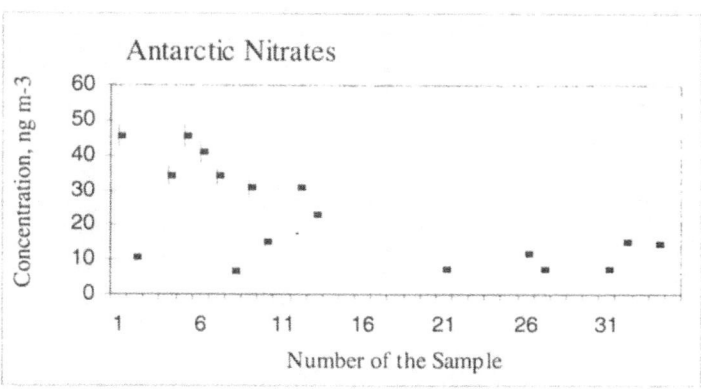

Figure 4. Concentration of anions, (chloride, sulfate and nitrate), present in PM-10 particulate matter collected sequentially from January (sample 1) to September 1996 on King George Island. The short vertical line indicates the standard deviation.

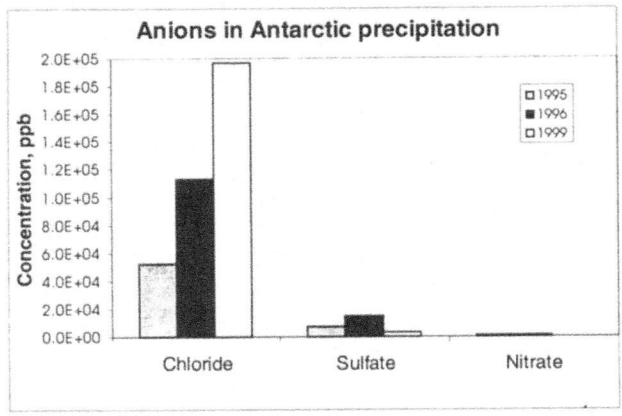

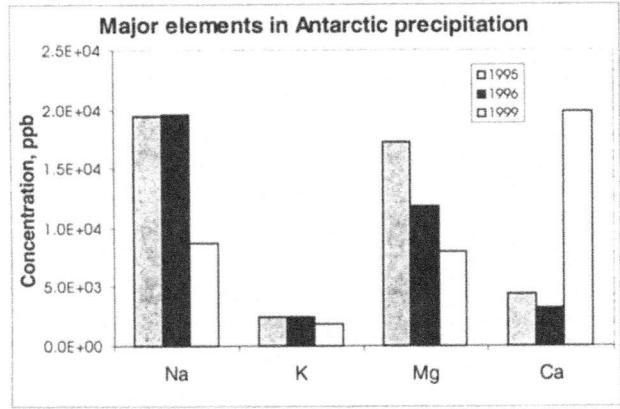

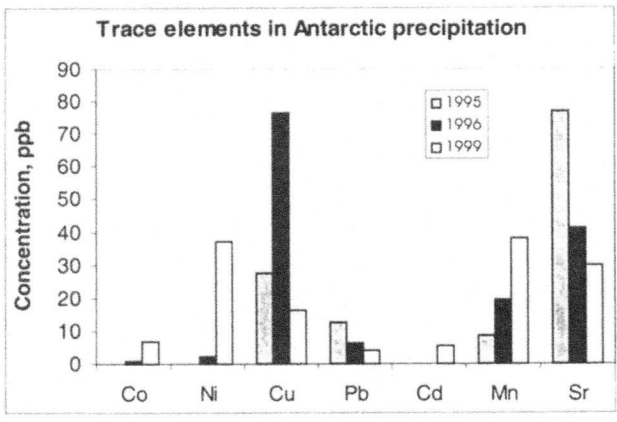

Figure 5. Mean annual concentrations of anions (chloride, sulfate and nitrate) and major trace elements (sodium, potassium, magnesium and calcium), in Antarctic precipitation for three different years, 1995, 1996, and 1999.

Elements can reach the water from the atmosphere (aerosols and precipitation) or from sediments removed by continual water movement by wind at surface banks or by deep water currents. They can be distributed in various ways, as particulate matter, or sediments. Figure 6 shows the concentration of six trace elements in the surface waters of three ponds called AGAB, BASU, and INCH, on King George Island (Préndez et al., 1999). We compare the total element content in water (particulate matter plus dissolved elements in water) and the element concentration in the column of water (dissolved elements), in order to estimate the particulate matter element concentration. It is clear that some of the elements, Ni, Cu, Se and Cd, in some ponds, are essentially concentrated in particulate matter and their presence is probably related to organic matter which forms complexes or chelates with the elements, thus extracting them from the column of water.

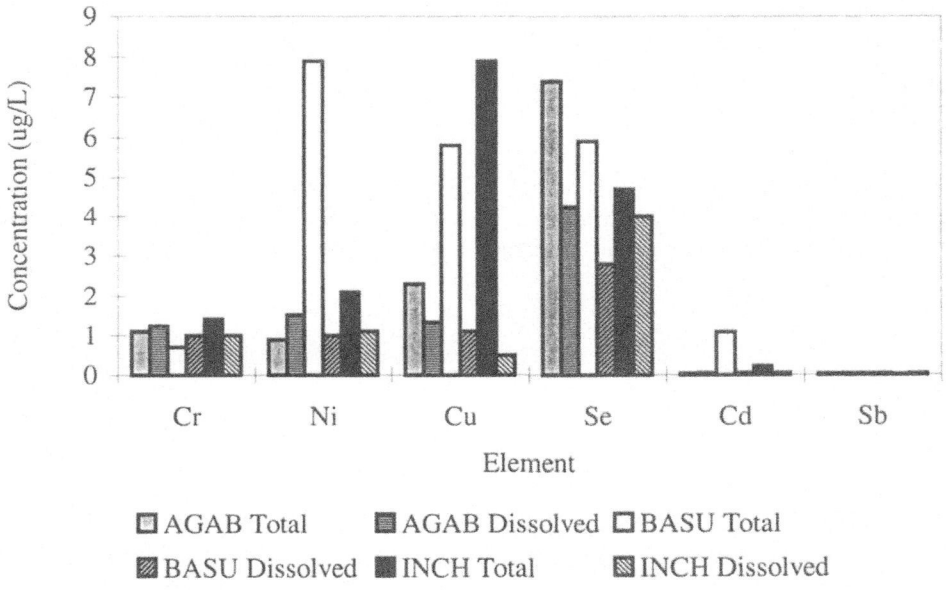

Figure 6. Mean concentration of six trace elements in surface waters of three ponds called AGAB, BASU, and INCH, on King George Island, Antarctic Peninsula.

Lichens are the most important autotrophic organisms in Antarctica, in terms of their diversity, distribution and biomass. They can absorb different chemical species directly from the air and live for long periods under very extreme conditions. Therefore, we thought it would be interesting to evaluate their behavior as biomonitors of atmospheric pollution. To do this, we used an extension of the enrichment factor (EF) method. This method compares the relative chemical content (relative to a standard, in lichens, in this case), to the chemical content in another part of the environment under study. EF values of around 1-5 indicate a natural origin; much greater EF values indicate a man-made origin. Table 4 shows the element composition of the Antarctic lichen *Neuropogon aurantiaco ater*. Values vary to a large extent with time, but not in any systematic way.

Table 4. Element composition of lichen *Neuropogon aurantiaco ater*, expressed as mg/kg for samples collected over a twenty-year period (1977 to 1997).

Year	Ca	Fe	Mg	Mn	Cd
1977	13500-17400	2750-3400	385-465	19.0-27.0	0.02-0.04
1983	24000-31000	2450-4050	355-465	7.8-11.2	0.02-0.04
1989	7900-8100	1400-2200	535-605	6.6-9.4	0.034-0.036
1996	9500-10500	1650-3850	450-550	13.6-18.7	0.02-0.03
1997	10000-10500	2100-2900	105-125	5.6-6.3	0.023-0.025

Year	Co	Cr	Cu	Ni	Pb	Zn
1977	0.3-0.4	0.9-1.55	2.2-2.5	5.3-8.2	0.35-0.45	2.2-3.5
1983	0.2-0.3	1.0-1.6	1.4-2.0	4.4-7.7	0.2-0.65	1.2-1.6
1989	0.2-0.3	1.2-2.0	0.5-0.85	2.5-4.8	0.15-0.35	1.4-1.9
1996	0.14-0.21	0.6-0.9	1.4-2.1	3.3-6.9	0.7-1.4	6.7-7.4
1997	0.07-0.10	1.0-1.4	0.9-1.35	4.2-7.0	1.2-1.3	5.9-6.0

Figure 7 applies the EF methodology to lichen collected during years 1983, 1989, and 1997. EF values for lead (Pb), normalized with respect to Mg concentrations from 1977, indicate an increasing presence of lead from man-made sources in the Antarctic air (Préndez *et al*, 1999).

Figure 7. Enrichment Factor (EF) for lichen *Neuropogon aurantiaco ater* collected during years 1983, 1995 and 1997. Lichen magnesium concentration is used as reference.

5. CONCLUSIONS

The study of the concentration, size, and composition of particulate matter is a good approach for learning about some aspects of atmospheric chemistry and their modification by either natural or man-made factors.

The study would be more complete if other parts of the environment, such as precipitation, soil, waters, or vegetation are simultaneously studied. In this way, a better description of the atmosphere over a certain period of time can be obtained.

The methodology and the appropriate detection limits obtained using ICP-MS and IC to analyze waters and atmospheric aerosols, including physical and chemical properties of particulate matter, justify the possible application of similar procedures to samples of fresh snow and ice from ice cores, though some technical adjustments may be needed, such as the sample injection procedure. These media show many similarities, such as only requiring very small sample sizes and very low detection limits for the analytical instrumentation. Thus, for example, a volcanic eruption could be characterized within an ice core, providing valuable information about past events and paleoclimate.

6. ACKNOWLEDGMENTS

INACH (Instituto Antártico Chileno) and the Antarctic Program of Universidad de Chile have financed this twenty-year period of research. I also thank all of the enthusiastic students who have participated in our work over the years.

7. REFERENCES

Aponte, R. A., 1998, Estudio integrado de variables químicas y físicas y sus interrelaciones para observar el cambio ambiental en la Península Antártica, Tesis de Magister en Química (M.Sc. Chem.), Universidad de Chile (in Spanish).

Boutron, C. F., and Wolf, E. W., 1989, Heavy metal and sulfur emissions to the atmosphere from human activities in Antarctica, *Atmospheric Environment*, **23(8)**:1669-1675.

Carrasco, M. A., and Préndez, M., 1991, Element distribution of some soils of continental Chile and the Antarctica Peninsula: projection to atmospheric pollution, *Journal of Water, Air, and Soil Pollution*, **57-58s**, 713-722.

Kerminen, V-M., Teinila, K., and Hillamo, R., 2000, Chemistry of sea-salt particles in the summer Antarctic atmosphere, *Atmospheric Environment*, **34**:2817-2825.

Mason, B., 1966, *Principles of Geochemistry*, Third edition, J. Wiley and Sons, New York, pp. 45-46.

Préndez, M. M., and Alcota, C., 1999, Análisis de tendencias del material particulado fino en la baja tropósfera sub-antártica: comportamiento físico y químico, I Reunión Chilena de Investigaciones Antárticas, Agosto 1999, Santiago, Chile, *Revista Serie Científica INACH*, **49**:120-122 (in Spanish).

Préndez, M., Carrasco, M. A., and Alcota, C., 1999, Total and dissolved elements in superficial waters of Antarctic Peninsula, 5th International Conference on the Biogeochemistry of Trace Elements, Vienna, July 1999, Vol. **II**, pp. 543-544.

Troncoso, W. N., and Préndez, M. M., 1999, Evolución de la composición física y química de los aerosoles sub-antárticos pm-10, I Reunión Chilena de Investigaciones Antárticas, Agosto 1999, Santiago, Chile, *Revista Serie Científica INACH*, **49**:132-134 (in Spanish).

INDEX

cecs
SOUTHERN PATAGONIA ICEFIELD

IMAGE MOSAIC

Satellite image mosaic of Southern Patagonia Icefield of 14 January 1986. The mosaic was geolocated using the 1:250.000 scale Preliminary Maps of the Instituto Geográfico Militar, Chile. It is a color composite of band 1 (blue), band 4 (green) and band 5 (red).

The image was first published in Naruse and Aniya, 1992, Outline of Glacier Research Project in Patagonia, 1990, *Bulletin of Glacier Research*, **10**:31-38.

The annotated image is adapted from Casassa, G., Rivera, A., Aniya, M., and Naruse, R., 2000, *Características glaciológicas del Campo de Hielo Patagónico Sur, Anales del Instituto de la Patagonia. Serie Ciencias Naturales*, **28**:5-22. Printed with permission from Anales, Instituto de la Patagonia.

This image is a large print of Fig. 3, Chapter 7, Casassa, G., Rivera, A., Aniya, M., and Naruse, R., 2002, Current knowledge of the Southern Patagonia Icefield, in: *The Patagonian Icefields: a unique natural laboratory for environmental and climate change studies*, eds., G. Casassa, F. V. Sepúlveda and R. M. Sinclair, Kluwer Academic / Plenum Publishers, New York, pp. 67-83.

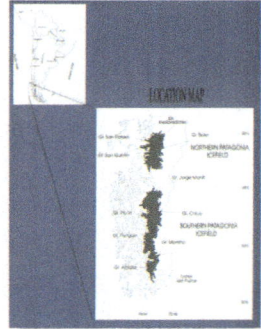

SOUTHERN PATAGONIA ICEFIELD

The Southern Patagonia Icefield (SPI) extends north-south for 370 km, between 48°15' S and 51° 35' S, at an average longitude of 73°30' W. It has an area of 13,000 km², a mean width of 35 km and a minimum width of 9 km. The first detailed glacier inventory was compiled by Aniya et al. (1996), who showed that the SPI is composed of 48 major outlet glaciers and over 100 small cirque and valley glaciers. These glaciers flow from the Patagonian Andes to the east and west, generally terminating with calving fronts in freshwater lakes (east) and Pacific Ocean fjords (west).

Most of the glaciers have been retreating with a few in a state of equilibrium and advance. Glacier retreat is interpreted primarily as a response to regional atmospheric warming and to a lesser extent, to precipitation decrease observed during the last century in this region.

CENTRO DE ESTUDIOS CIENTIFICOS

Avenida Arturo Prat 514, Casilla 1469, Valdivia, Chile
Tel.: +56 63 234500 Fax: +56 63 234517
Email: cecsvaldivia@cecs.cl www.cecs.cl

73° W

74° W

49°S · Villa O'Higgins

Bernardo

Témpano
O'Occidental

O'Higgins

Volcán Lautaro

Puerto Edén

49°S

Chico

Río N

HPS10

HPS12

Cº Murallón

HPS13

HPS18

HPS19

Penguin

Amalia

Europa

Upsala

Lago Argentino

50°S

Guilardi

Balmaceda

74° W

51°S

73° W

51°S

Puerto Natales

SCALE 1:500,000

0   20   40 km

Universal Transverse Mercator Grid UTM,
zones 18 and 19. Datum WGS84.

The manufacturer's authorised representative in the EU is Springer
Nature Customer Service Centre GmbH, Europaplatz 3, 69115 Heidelberg,
Germany. If you have any concerns regarding our products, please
contact ProductSafety@springernature.com

Printed and bound by CPI Group (UK) Ltd, Croydon, CR0 4YY
23/04/2026
02095585-0019